STILL IMAGE COMPRESSION ON PARALLEL COMPUTER ARCHITECTURES

THE KLUWER INTERNATIONAL SERIES IN ENGINEERING AND COMPUTER SCIENCE

MULTIMEDIA SYSTEMS AND APPLICATIONS

Consulting Editor

Borko Furht
Florida Atlantic University

Recently Published Titles:

INTERACTIVE VIDEO-ON-DEMAND SYSTEMS: ***Resource Management and Scheduling Strategies***, by T. P. Jimmy To and Babak Hamidzadeh
ISBN: 0-7923-8320-6

MULTIMEDIA TECHNOLOGIES AND APPLICATIONS FOR THE 21st CENTURY: ***Visions of World Experts***, by Borko Furht
ISBN: 0-7923-8074-6

INTELLIGENT IMAGE DATABASES: ***Towards Advanced Image Retrieval***, by Yihong Gong
ISBN: 0-7923-8015-0

BUFFERING TECHNIQUES FOR DELIVERY OF COMPRESSED VIDEO IN VIDEO-ON-DEMAND SYSTEMS, by Wu-chi Feng
ISBN: 0-7923-9998-6

HUMAN FACE RECOGNITION USING THIRD-ORDER SYNTHETIC NEURAL NETWORKS, by Okechukwu A. Uwechue, and Abhijit S. Pandya
ISBN: 0-7923-9957-9

MULTIMEDIA INFORMATION SYSTEMS, by Marios C. Angelides and Schahram Dustdar
ISBN: 0-7923-9915-3

MOTION ESTIMATION ALGORITHMS FOR VIDEO COMPRESSION, by Borko Furht, Joshua Greenberg and Raymond Westwater
ISBN: 0-7923-9793-2

VIDEO DATA COMPRESSION FOR MULTIMEDIA COMPUTING, edited by Hua Harry Li, Shan Sun, Haluk Derin
ISBN: 0-7923-9790-8

REAL-TIME VIDEO COMPRESSION: ***Techniques and Algorithms***, by Raymond Westwater and Borko Furht
ISBN: 0-7923-9787-8

MULTIMEDIA DATABASE MANAGEMENT SYSTEMS, by B. Prabhakaran
ISBN: 0-7923-9784-3

MULTIMEDIA TOOLS AND APPLICATIONS, edited by Borko Furht
ISBN: 0-7923-9721-5

MULTIMEDIA SYSTEMS AND TECHNIQUES, edited by Borko Furht
ISBN: 0-7923-9683-9

STILL IMAGE COMPRESSION ON PARALLEL COMPUTER ARCHITECTURES

by

Savitri Bevinakoppa
Royal Melbourne Institute of Technology
Melbourne, Australia

SPRINGER SCIENCE+BUSINESS MEDIA, LLC

ISBN 978-0-7923-8322-2 ISBN 978-1-4615-4967-3 (eBook)
DOI 10.1007/978-1-4615-4967-3

Library of Congress Cataloging-in-Publication Data

A C.I.P. Catalogue record for this book is available
from the Library of Congress.

Printed on acid-free paper.

To

M egha

M anisha

M ayank

CONTENTS

LIST OF FIGURES

LIST OF TABLES

LIST OF NOTATIONS

Notations	Description
η	Efficiency
BBIP	Block Based Image Partitioning
BCF	Block Compression Factor
BCT	Block Comparator Technique
BD	Block Dependency
BIV	Block Intensity Value
BN	Block Number
BWIP	Balanced Workload Image Partitioning
CDL	Component Distribution Language
CIDS	Compressed Image Data Structure
DCT	Discrete Cosine Transform
DCuT	Distributed Memory Architecture with Cube Topology
DMA	Distributed Memory Architecture
DPyT	Distributed Memory Architecture with Pyramid Topology
DToT	Distributed Memory Architecture with Torus Topology
DTrT	Distributed Memory Architecture with Tree Topology
EIL	Equal Intensity List
EOI	End Of Image Marker
HMA	Hybrid Memory Architecture
IBD	Inter-Block Dependency
ICR	Image Compression Ratio
ICR_{BCT}	Image Compression Ratio for Block Comparator Technique
ICRIF	Image Compression Ratio Improvement Factor
ICR_{NBCT}	Image Compression Ratio (ICR) for Non-Block Comparator Technique
ICT	Image Compression Technique used for image processing
IPC	Inter-Processor Communication
IPM	Image Partitioning Method
JPEG	Joint Photographic Experts Group
MA	Memory Architecture
MCU	Minimum Coded Unit
ML	Match List
MO	Memory Organisation

MPEG	Motion Photographic Experts Group
MPP	Message Passing Primitives
NB	Number of Blocks
NBCT	Non-Block Comparator Technique
N_{BEIL}	Number of blocks in Equal Intensity List
NIBD	Non-Inter-Block Dependency
NIPC	Non-Inter-Processors Communication
NL	Number of Similar Block Lists in SBG
nl	Number of Block Numbers in each Unique Block list
NP	Number of processors
NSB	Number of Similar Blocks
NT	Network Topology
NUB	Number of Unique Blocks
Px	Plan-x for implementation
P1	P(BCT, NIPC, BWIP, DMA, DTrT, NP)
P2	P(NBCT, NIPC, BWIP, SMA, SGM, NP)
P3	P(NBCT, NIPC, BWIP, SMA, SLgM, NP)
P4	P(NBCT, NIPC, BWIP, DMA, DTrT, NP)
P5	P(NBCT, NIPC, BWIP, DMA, DToT, NP)
P6	P(NBCT, NIPC, BWIP, DMA, DPyT, NP)
P7	P(NBCT, NIPC, BWIP, DMA, DCuT, NP)
P8	P(BCT, NIPC, BWIP, SMA, SGM, NP)
P9	P(BCT, NIPC, BWIP, SMA, SLgM, NP)
P10	P(BCT, NIPC, BWIP, DMA,DToT, NP)
P11	P(BCT, NIPC, BWIP, DMA, DPyT, NP)
P12	P(BCT, NIPC, BWIP, DMA, DCuT, NP)
P13	P(BCT, IPC, BWIP, SMA, SlgM, NP)
P14	P(BCT, IPC, BWIP, DMA, DToT, NP)
P15	P(BCT, IPC, BWIP, DMA, DPyT, NP)
PE	Processing Element
POSIC	Portable Operating Set of Instruction Codes
RL	Reference List
S	Speedup for NP processors
S_{BCT}	BCT Compressed image size in Bytes
S_{BCT1}	BCT Compressed image size for CIDS-1 in Bytes
SBG	Similar Block Group
SBL	Similar Block Lists
S_{Blk}	Size of one block in Bytes
S_{BNF}	Size of the Block Number Field
$S_{Compimg}$	Compressed image data size in Bytes
SD	Storage Device
S_{EOI}	Size of End Of Image marker
SIF	Speed Improvement Factor
S_{JHI}	Size of JPEG Header Information in Bytes
SLgM	Shared Memory Architecture with Local-Plus-Global

	Memory organisation
SGM	Shared Memory Architecture with Global Memory organisation
SMA	Shared Memory Architecture
S_{NBCT}	NBCT Compressed image size in Bytes
SOF	Start Of Frame
SOI	Start Of Image marker
SOS	Start Of Scan
S_{SBM}	Size of the Similar Block Marker
S_{Srcimg}	Source image data size in Bytes
S_{UBM}	Size of the Unique Block Marker
S_{UBNF}	Size of the Unique Block Number Field
T_1	Time taken by a single processor
T_{BC}	Total number of Base Operations required for Block Comparison
T_{CS}	Total number of Base operations required for subtraction operation for one 8 x 8 block
TD	Transfer Device
T_{DCT}	Total number of Base Operations required for DCT step for one 8 x 8 block
T_{dct}	Total number of Base operations required for DCT function for one 8 x 8 block
T_{Enco}	Total number of Base Operations required for Huffman Encoding for one block
$T_{Intcomp}$	Total number of Base Operations for block intensity value comparison
T_{JPEG}	Total number of Base Operations taken by the JPEG algorithm
T_N	Time taken by N processors
T_{Quan}	Total number of Base Operations required for Quantisation step for one 8 x 8 block
$T_{sampblock}$	Total number of Base Operations required for sample-by-sample comparison of one block
$T_{sampcomp}$	Total number of Base Operations for comparing samples of a block with those of existing Unique Blocks
$T_{sampsum}$	Total number of Base Operations required for the summation of sample values in any image block
T_{sum}	Total number of Base Operations for summation of samples in all image blocks
UBG	Unique Block Group
UBIV	Unique Block Intensity Value
UBN	Unique Block Number
VQ	Vector quantisation

PREFACE

This book explains investigation of the application of parallel processing techniques to digital image compression. Digital image compression is used to reduce the number of bits required to store an image in computer memory and/or transmit it over a communication link. Over the past decade advancements in technology have spawned many applications of digital imaging, such as photo videotex, desktop publishing, graphics arts, colour facsimile, newspaper wirephoto transmission, medical imaging. For many other contemporary applications, such as distributed multimedia systems rapid transmission of images is necessary. Dollar cost as well as time cost of transmission and storage tend to be directly proportional to the volume of data. Therefore, application of digital image compression techniques become necessary to minimise costs.

A number of digital image compression algorithms have been developed and standardised. With the success of these algorithms, research effort is now directed towards improving implementation techniques. Joint Photographic Experts Group (JPEG) and Motion Photographic Experts Group (MPEG) are international organisations which have developed digital image compression standards. Hardware (VLSI chips) which implement the JPEG image compression algorithm are available. Such hardware is specific to image compression only and can not be used for other image processing applications. A flexible means of implementing digital image compression algorithms is still required. An obvious method of processing different imaging applications on general purpose hardware platforms is to develop software implementations.

JPEG uses an 8 x 8 block of image samples as the basic element for compression. These blocks are processed sequentially. There is always a possibility of having similar blocks in a given image. If similar blocks in an image is located, then repeated compression of these blocks is not necessary. By locating similar blocks in the image, speed of compression can be increased and the size of compressed image can be reduced. Based on this concept an enhancement to the JPEG algorithm is proposed, called Block Comparator Technique (BCT).

Most of the current implementation of JPEG and MPEG compression methods are in sequential form. Parallel processors are becoming more affordable and are likely to be used quite extensively in the near future. Therefore various options for implementing digital image compression algorithms were investigated on parallel computer architectures.

ACKNOWLEDGMENTS

First and foremost I would like to express my appreciation and my sincere gratitude to Ass. Prof. Nalin K. Sharda and Dr. Hema Sharda. Their inspiration, enthusiasm and encouragement have made this book successful.

I wish to thank my husband Gangadhar, my daughters Megha, Manisha and my son Mayank for their love and support over the years. My special thanks to my father V. B. Nandi who encouraged me to undertake this research which was both challenging and rewarding.

I would like to thank the Department of Computer System Engineering, RMIT, Melbourne for providing the Mercury system, Defence Science and Technology Organisation (DSTO), Adelaide for their contribution to the Shiva system installed in the Department of Computer and Mathematical Sciences, Victoria University of Technology. I would also like to thank Dr. Vijay Bhatkar, Sampath, Suhas and other staff members at Centre of Development of Advanced Technology (C-DAC), Pune, India for their support and assistance carrying out the comparitive studies on the Param supercomputer.

I would like to extend my sincere thanks to Ass. Prof. Peter Cerone and my collegues, who proof read this manuscript and made useful corrections.

1 INTRODUCTION

1.1 Introduction

This chapter gives an introduction to the existing digital image compression techniques, parallel processing techniques, the research problem and investigation procedures described in this book.

A literature survey was undertaken to study the existing digital image compression techniques, performance improvement techniques, and parallel processing techniques. The Joint Photographic Experts Group (JPEG) algorithm was selected for this research. At present JPEG standard compression process is done block-by-block in a sequential manner. An enhancement to the current JPEG compression technique is proposed. The aims of this enhancement are to speedup the operation and reduce the compressed image size. Implementation of the JPEG algorithm on parallel computers, to further speedup compression operations, has also been studied.

Digital image compression is used to reduce the number of bits required to store an image in computer memory and/or transmit it over a communication link [Jain, 89]. Image compression prior to transmission should reduce the amount of information to be transmitted, thus lowering the bandwidth requirements and cost. The main focus of this research is to enhance the performance of the current digital image compression method. Details of the research problem are given in section 1.2.

A literature review of existing digital image compression standards, and techniques to improve compression parameters such as quality, speedup and compression ratio are discussed in section 1.3.

Research objectives are explained in section 1.4. Outline of book chapters is given in section 1.5.

1.2 Problem Statement

Transmission of image data using simple techniques requires a bit rate that is too large for many communications links or storage devices. Digitisation may be desirable for security and/or reliability, but it can cause bandwidth explosion. Hence data compression is required to use the available bandwidth as effectively as possible.

Over the past decade advancements in technology have spawned many applications of digital imaging, such as photo videotex, desktop publishing, graphics arts, colour facsimile, newspaper wirephoto transmission, medical imaging. For many other contemporary applications, such as distributed multimedia systems rapid transmission of images is necessary. Images are used in multimedia for browsing, retrieval, storage and slide show. Research challenge includes developing real-time compression algorithms and guaranteed Quality of Service in multimedia applications [Furht, 94] [Furht, 95]. Dollar cost as well as time cost of transmission and storage tend to be directly proportional to the volume of data. Therefore, application of digital image compression techniques become necessary to minimise these costs.

A number of digital image compression algorithms have been developed [Aravind, 89] and standardised, such as the JPEG, the MPEG and PX64 standards. Most of the current implementations of JPEG and MPEG compression methods are in sequential form. Parallel processors are becoming more affordable and are likely to be used quite extensively in the near future. Thus techniques for parallel processing of image compression can deliver substantial dividends. In this book an improvement to the JPEG algorithm and a study of techniques for parallel implementation of image compression is presented.

1.3 Literature Review

Literature survey covered similar work reported in journals and conference proceedings. To provide an overview of previous work and to provide a basic theoretical understanding of the subject, the papers presented by various authors are reviewed and quoted in this chapter. The areas covered in the literature survey are: digital image compression techniques, standards such as JPEG, MPEG, PX64, parallel implementations, performance analysis issues.

Digital image compression techniques are discussed in section 1.3.1. Image compression algorithms include optimisation of parameters, such as quality, complexity, compression ratio and speedup of operation. Techniques employed to improve these parameters are discussed in section 1.3.2.

1.3.1 Digital Image Compression Techniques

Various digital image compression techniques, hardware, and software are discussed in this section.

Borko Furht has presented a classification of digital image compression technique in [Furht, 92]. Digital image compression techniques can be broadly classified into still image compression and motion image compression techniques. Still image compression techniques can be further classified into lossy compression and lossless compression techniques. Lossless compression techniques are used to recover the original image representation perfectly, whereas a lossy compression technique is used to output image similar to the original one. Lossy compression provides higher compression ratio. Lossless digital image compression techniques can be classified based on encoding technique such as Huffman coding, Arithmetic decomposition, Lempel Ziv, and Run length. Lossy compression techniques are classified into prediction based technique, frequency oriented techniques, importance oriented techniques, and hybrid techniques [Furht, 95]. Prediction based techniques predict subsequent values by observing previous values. Frequency oriented technique apply the Discrete Cosine Transform (DCT). Importance oriented techniques use some important characteristics of images as the basis for compression. The hybrid compression techniques, such as JPEG, MPEG and PX64 use several approaches such as DCT, Vector Quantisation, prediction technique.

Digital image compression techniques can also be classified based on the algorithms used such as Wavelet transform, Fractal image, Vector Quantisation and DCT. The Wavelet transform algorithm is based on basis functions [Koornwinder, 93]. Fractal images are based on Iterated Function Systems (IFS) [Barnsley, 93]. Vector Quantisation is based on vector representation of the image and based on code book design [Cosman, 96] [Gersho, 92]. The JPEG algorithm is based on Differential Pulse Code Modulation (DPCM) and the DCT [Pennebaker, 93].

Wavelet transform can also be used with the JPEG standard in the video industry for on-line editing [Cornell, 93]. The VQ method is complicated by the need for code design. Therefore, coding with Vector Quantisation is slow as compared to coding with the JPEG algorithm. VQ is more efficient when it is combined with other techniques. The JPEG standard is widely used for still imaging applications. The JPEG algorithm is used in the standard developed by the Motion Pictures Expert Group (MPEG), for compressing moving pictures as well. Therefore, the JPEG algorithm was chosen for this research purpose.

Aravind has described a number of digital image compression algorithms and standards, such as the JPEG, the MPEG and PX64 [Aravind, 89]. The JPEG standard is described in sufficient detail in [Nelson, 92a], [Pennebaker, 93] and [Wallace, 92]. A very succinct description of the various techniques used in the JPEG standard is given by William Pennebaker in [Pennebaker, 93]. The MPEG standard is described in [Gall, 91] and [Draft, 90]. The PX64 compression algorithm for video telecommunications is described in [Liou, 91]. PX64 algorithm consists of DCT-based intraframe compression, which is similar to JPEG algorithm and predictive interframe coding based on Differential Pulse Code Modulation (DPCM) and motion estimation. Therefore all these standards use the DCT-based method of compression as a basic step.

Two prominent image compression techniques are predictive technique and DCT-based technique [Pennebaker, 93]. The JPEG was working on still image compression using both techniques. The predictive technique is a lossless compression technique while the DCT - based technique is a lossy technique. The DCT-based method of compression is widely used, as it is suitable for a large number of applications, and also, it is expected that DCT-based technique developed for implementing the JPEG standard can be applied to compressing motion pictures as well; because the MPEG standard is also based on the DCT .

Some of the hardware chips for digital image compression in VLSI implementation are Toshiba's VLSI processor T9506 [Sugai, 87], C-Cube's JPEG CL550 chipset, SGS-Thomson's ST1140 CMOS chip [Leonard, 91], and Intel's Digital Video Interactive (DVI) chip [Vaaben, 91] i750 video processor [Harney, 91].

1.3.2 Performance Improvement

In developing digital coders many parameters need to be considered, such as bit rate, quality of output image, complexity of the algorithm, compression ratio, quality of service and speed of operation. Reduced bit rate reduces quality, unless complexity of the coding technique is increased. Complexity raises cost, and in many coding techniques it increases the processing delay as well.

The JPEG algorithm compresses the image based upon a user specified quality factor, where for higher quality of output image lower compression ratio can be achieved and vice verse. In the JPEG compressed data structure block numbers are not specified. If any block is lost during transmission then the output image is not the same as the input image.

Papathanassiadis T. [Papathanassiadis, 92] discussed compressed image data structure with block numbers. This has the potential of improving the quality of service. But, by including block numbers the compression ratio gets reduced.

Roberto Rinaldo [Rinaldo, 95] has discussed block matching technique for fractal image coding technique. The proposed coding scheme consists of predicting blocks in one subimage from blocks in lower resolution subbands with the same orientation. This block prediction scheme is simpler than the iterative scheme adopted in standard fractal block coders and visual quality is better than the other schemes. A drawback of Rinold's scheme is the larger encoding time required in comparison to the time required in coding techniques like JPEG.

The DCT-based methods work on each block of image independently, therefore, the JPEG algorithm can be parallelised by processing each image block on a separate processor. The JPEG algorithm can thus be implemented on parallel computer architectures.

Rapid advances in electronics technology throughout the 1980s has allowed more complex, yet relatively inexpensive computational devices with greatly increased throughput to be developed. New concurrent (or parallel) techniques using fast sequential processing devices, and multi-processing devices are now being applied to digital data compression. Existing parallel implementations of digital image compression are discussed below.

The Digital Video Interactive (DVI) algorithm was implemented on the MEiKO and the iPSC/2 parallel architectures [Tinker, 89]. The MEiKO computer is based on the T414 transputer. It comprises 65 transputers, and the software is written in

OCCAM and C programming language. The iPSC/2 is a hypercube parallel computer based on Intel's i80386 microprocessor. The compression algorithm on a 64 node MEiKO computer took 13.85 sec/frame and on a 64 node iPSC/2 computer it took 9.05 sec/frame. Therefore, on the 64 node MEiKO or the iPSC/2 computers compression algorithm could not achieve real-time compression. Even if the number of processors is increased in the iPSC/2 computer, the minimum compression time that could be obtained is nearly 2 sec/frame [Tinker, 89].

Compact Disc-Interactive (CD-I) full motion video encoding algorithm was implemented on Parallel Object Oriented Machine (POOMA). This system was developed at the Philips research laboratories. It is based on the Motorola MC68020 with a loosely coupled MIMD architecture and consists of 100 nodes. Compression algorithm took less than 2 sec/frame on 100-processor nodes [Sijstermans, 91]. For the parallel algorithm used, saturation will occur if more than 100 processors are used. Thus, for real-time applications even this system is not quite adequate.

The HDTV Codec is based on a motion-adaptive DCT algorithm. It consists of a parallel signal processing architecture and LSI gate array [Kinoshita, 92]. This hardware compresses the motion picture at the bit rate of 130 Mb/s, that is, in real-time. This hardware is specific to motion image compression.

John Elliott [Elliott, 89] describes simulation of image compression algorithm on a supercomputer based on the Transputer processor along with the architecture of the Edinberg Concurrent supercomputer. The parallel algorithm used on this supercomputer can process 6 - 7 frames/sec by optimising the code. But for real-time image compression a speed of at least 18 - 20 frames /sec is required.

M. N. Chong [Chong, 90] describes implementation of the adaptive transform coding technique on a transputer based quadtree architecture. There is a limitation to the degree of parallelism that can be achieved in this implementation. The results obtained on the quadtree structure for various sized networks are given in this paper. The least execution time of 1.538 sec. is obtained on 16 processors. This execution time is higher than that required for real-time image compression.

R. Aravind [Aravind, 89] explains implementation of the DCT-based JPEG decompression algorithm on a Digital Signal Processor (DSP)-based system. This decoder is capable of processing in real-time, at approximately 15 frames/sec with a frame size of 128 x 96.

Srinath Ramaswamy [Ramaswamy, 93] describes a parallel pipelined DSP-based architecture for implementing the DCT-based JPEG algorithm with arithmetic coding. He has given the experimental results of executing the JPEG algorithm on a DSP-based architecture for a 256 x 256 pixel monochrome still image. The execution time varies from 0.61 sec. to 0.12 sec as the number of processors is increased from one to six. For a large image size, image compression can be achieved in close to real-time by increasing the number of DSP processors in the network.

Placement of blocks of image data on different parallel architectures is one of the many issues that was explored and investigated further. Papathanassiadis T. [Papathanassiadis, 92] discussed various image partitioning strategies. There are two main methods used for image partitioning: with interblock dependency and without interblock dependency. Chung-Ta King [Chung-Ta King, 91] discussed strategies for partitioning and processing images with interblock dependency on distributed memory multi-computers. Browne [Browne, 89] discussed the various options of image processing mapping methods onto Transputer networks.

1.4 Research Objectives

JPEG is one of the most widely used image compression standard. This research is focused on improving the performance of this standard, and its implementation on parallel architectures. Hardware (VLSI chips) which implement the JPEG image compression algorithm are available. Such hardware is specific to image compression only and can not be used for other image processing applications. A flexible means of implementing digital image compression algorithms is still required. An obvious method of processing different imaging applications on general purpose hardware platforms is to develop software implementations.

JPEG uses an 8 x 8 block of image samples as the basic element for compression. These blocks are processed sequentially. There is always a possibility of having similar blocks in a given image. If the similar blocks in an image are located, then repeated compression of these blocks is not necessary. By locating similar blocks in the image, speed of compression can be increased and the size of compressed image can be reduced. Based on this concept an enhancement to the JPEG algorithm, called the Block Comparator Technique (BCT) is proposed. For many applications rapid transmission of images in real-time and good quality of service is required. Various options for enhancing the current JPEG standard is investigated, to reduce the compressed image size and to improve the speed of compression.

One of the primary objectives of this research project was to develop techniques for exploiting parallel processing systems for real-time image compression and decompression. It is expected that such parallel processing technique will not only reduce the execution time, but will also accomplish other significant performance improvements such as improved quality of compressed image, improved reliability and availability of the system, and better scalability. Therefore various options are investigated for implementing digital image compression algorithms on parallel architectures.

Some of the implementation options were studied by simulating these on computer models. A simulation package called NETWORK II.5 was used for building the computer model and running the required experiments on the same. Simulation results were used to determine speedup, scaleup and efficiency of the techniques developed.

1.5 Book Outline

This section gives a brief description of each of the following chapters.

Chapter 2 Digital image compression techniques: In this chapter different digital image compression techniques, and the JPEG image compression standard are described. Digital image compression techniques are based on algorithms such as Wavelet transform, Fractal images, Vector Quantisation and Discrete Cosine

Transform. Digital image compression technique developed by the Joint Photographic Experts Group is based on the Discrete Cosine Transform.

The JPEG technique is applicable to a wide variety of applications and is one of the most widely used technique. Therefore, JPEG technique is chosen as the main focus for our research. Present JPEG compression process is done block-by-block in a sequential manner. An enhancement to the current JPEG compression technique is proposed, to speedup the operation and reduce the compressed image size.

Chapter 3 Parallel processing plans for digital image compression techniques: This chapter describes methods used for parallel processing of digital image compression algorithms. Types of parallel computers and parallel processing 'Plans' for digital image compression are described. Parallel computers are classified based on memory access technique, network topology and some other issues.

Digital image compression can be performed on parallel computers in a variety of ways. Each uniquely identifiable way of implementation is called a Plan. Each Plan can be specified as a 6-tuple consisting of image compression technique, block dependency, image partitioning method, memory architecture, network topology and the number of processors. Some of these Plans were implemented on available parallel computers and other Plans were simulated using the Network II.5 simulation package.

Model building and simulation involves ten steps, viz. problem formulation, model building, data collection, model translation, model verification, model validation, experiment planning, experimentation, analysis of results, and documentation. Each of these steps are described briefly in this chapter.

Chapter 4 Implementation of the JPEG algorithm on parallel computers: This chapter describes the hardware architecture and methods used for the implementation of the JPEG algorithm on parallel computer systems such as Mercury, Shiva and Param. The Mercury system has a distributed memory architecture. Shiva system has a shared memory architecture, and the Param system uses hybrid memory architecture.

JPEG algorithm was implemented on these three parallel computers with different image sizes and on various sized networks. This chapter describes implementation of the JPEG algorithm on three parallel computer systems and it gives the experimental results obtained on the same.

Chapter 5 Simulation of digital image compression techniques: This chapter describes modelling and simulation methods used for investigating parallel processing of image compression techniques, using the Network II.5 simulation package. Image compression Plans have been modelled for different parallel computer architectures using the Network II.5 simulation package. This chapter describes details of the model building process and the process of running simulation experiments for various Plans. Simulation results for these Plans are compiled to evaluate the performance of these Plans.

Speedup, scaleup and efficiency obtained for each Plan is given and the performance of different Plans are compared.

Chapter 6 Conclusions and future research: This chapter gives the conclusions and directions for future research. The Block Comparator Technique as well as parallel implementation aspects are discussed.

CHAPTER 2 DIGITAL IMAGE COMPRESSION TECHNIQUES

2.1 Introduction

This chapter describes digital image compression techniques, and the JPEG image compression standard. Digital image compression techniques are based on algorithms such as the Wavelet transform, Fractal images, Vector Quantisation and Discrete Cosine Transform (DCT). The digital image compression technique developed by the Joint Photographic Experts Group (JPEG) is mainly based on the quantisation of the DCT.

The JPEG technique is applicable to a wide variety of applications and is one of the most widely used technique. Therefore, the JPEG technique is chosen as the main focus for this research. Presently, JPEG compression process is done block by block in a sequential manner. An enhancement to the current JPEG compression technique is proposed. The aim of this enhancement is to speedup the operation and reduce the compressed image size.

The JPEG algorithm can be implemented on parallel computers to further speedup the compression and decompression operations.

Digital image compression techniques can be broadly classified into still image compression and motion image compression techniques. Still image compression techniques can be further classified based on the algorithm used for compressing the image such as Wavelet transform, Fractal images, Vector Quantisation (VQ) and the Discrete Cosine Transform (DCT). These digital image compression algorithms are described in section 2.2.

The JPEG standard is widely used for still imaging applications. The JPEG algorithm is used in the standard developed by the Motion Pictures Expert Group (MPEG), for compressing moving pictures as well. Therefore, The JPEG algorithm was chosen for our research. Section 2.3 describes the JPEG algorithm in detail.

JPEG uses an 8 x 8 block of image samples as the basic element for compression. These blocks are processed sequentially. There is always a possibility of having similar blocks in a given image. If similar blocks are located in an image, then repeated compression of these blocks is not necessary. By locating similar blocks in an image, speed of compression can be increased and the size of compressed image can be reduced. The technique used to enhance the JPEG algorithm is called Block

Comparator Technique in this book. This Block Comparator Technique (BCT) is described in section 2.4.

2.2 Digital Image Compression Techniques

By using mathematical methods such as Fourier transform, it is possible to represent a given image in terms of a few basis functions [Hunt, 93]. Recently, mathematicians, scientists and engineers have been active in seeking new methods for representing signals or data in terms of basis functions. Because these functions can be analysed, understood and characterised in a succinct manner, these methods can be applied to digital image compression and many other applications.

Based on the mathematical methods used in digital image compression, still image compression techniques can be classified as lossy compression techniques or lossless compression techniques. A classification tree for digital image compression techniques is shown in figure 2.1. Lossy compression techniques can compress the image down to 50 : 1 ratio, where-as lossless compression techniques can compress the image only upto a ratio of 3:1. The lossy compression technique can be further classified based on the algorithm used such as Wavelet transform, Fractal images, Vector Quantisation and DCT. Lossless compression technique include the Joint Bi-level Image Experts Group (JBIG) algorithm and the JPEG algorithm [Pennebaker, 93]. Lossy compression techniques are used in applications such as colour facsimile, newspaper wire-photo transmission, medical imaging, graphics arts, photovideotex, desktop publishing, and many other still imaging applications.

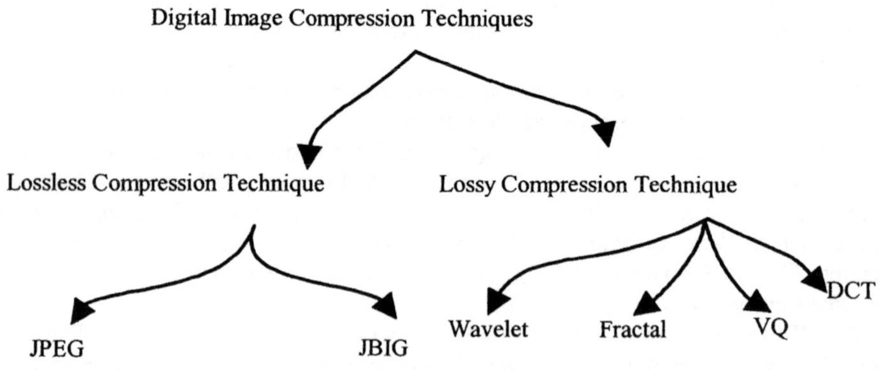

Figure 2.1 Digital image compression techniques

The Wavelet transform algorithm is based on basis functions; these are described in section 2.2.1. Fractal images are based on Iterated Function Systems (IFS); these are described in section 2.2.2. Vector Quantisation is based on vector representation of the image and code book design; this is described in section 2.2.3. The JPEG algorithm is based on the DCT and Differential Pulse Code Modulation (DPCM). The DCT-based method is described in section 2.2.4.

2.2.1 Wavelet Transform

There are two types of Wavelet Transforms ie. Continuous Wavelet Transform and Discrete Wavelet Transform. The Continuous Wavelet Transform was first presented by Grossmann and Morlet in 1984. Thereafter it was developed by others, including Holschneider (1988), Arneo'odo et al. (1989), Forge (1992). Daubechies (1986 and 88) was one of the first to work on Discrete Wavelet Transform. Wavelet transformation has a number of applications in signal processing and data compression.

Wavelet transform breaks the signal into a number of wave pulses (wavelets) that can be dilated and translated in two or more dimensions; and if the wavelet is anisotropic, it can also be rotated. These wave pulses are represented in terms of an amplitude function and can be analysed using scale and position of a signal [Hunt, 93]. By choosing an appropriate wavelet one can look at the properties of a signal, such as, amplitude and time scale of the original signal.

Wavelet basis functions are orthonormal. Therefore, these transformations can be used to remove redundant signals from the original signal, this leads to compression of the original signal.

In Wavelet compression the original multi-resolution image is decomposed into a low resolution signal and a difference signal [Nacken, 93] [Koornwinder, 93]. The low resolution signal is an average of the low frequency signals and is calculated by applying low pass filtering, followed by subsampling. The low resolution signal can be described by a smaller number of samples than the original image. The difference signal is the difference between the low resolution image and the actual image. The difference signal can be coded with a smaller number of bits per pixel. Thus the total number of bits required to encode the image is smaller than the original image.

2.2.2 Fractal Image Compression

Fractal image compression is based on Iterated Function System (IFS) theory and the Collage theorem. Fractal image compression can be achieved via the IFS compression algorithm, which is an interactive image modelling method based on the Collage theorem.

IFS fractals can be obtained through suitable transformation of the image [Barnsley, 93]. Such fractals can be used as approximates for real world images. Real world image is one of many basic shapes, such as a leaf or a letter of the alphabet, or a black and white fern, or a black cat sitting in a field of snow, etc. These fractals have the property that they are themselves models for real world images, and at the same time can be defined by finite strings of zeros and ones. This makes them suitable for image compression.

The Collage theorem says that "to find an IFS whose attractor is close to looks like a given set, one must try to find a set of transformations, contraction mappings on a suitable space within which the given set lies, such that the union, or collage, of the images of the given set under the transformations is close to or looks like the given set" [Barnsley, 93]. The degree to which two images look alike is measured using the Hausdorff metric.

2.2.3 Vector Quantisation

In the Vector Quantisation (VQ) technique image signals are represented by vectors of samples. Vector Quantisation can be viewed as a form of pattern recognition where an input pattern is approximated by one of a predetermined set of standard patterns. That is, the input signal is vector quantised in such a way that the input pattern is matched with one of a stored set of templates or codewords. In this way the complicated image signal can be replaced by a series of simple table lookups.

The following theorem shows that VQ can at least match the performance of any arbitrarily given coding system that operates on a vector of signal samples or parameters. Theorem is as follows. " *For any given coding system that maps a signal vector into one of N binary words and reconstructs the approximate vector from the binary word, there exists a vector quantiser with codebook size N that gives exactly the same performance, ie. for any input vector it produces the same reproduction as the given coding system*" [Gersho, 92].

Pattern matching with set of codebooks is done in several ways, such as nearest neighbour quantiser and exhaustic search algorithm [Gersho, 92]. In the nearest neighbour quantiser search algorithm, a vector is represented by the nearest vector stored in the codebook. An advantage of such an encoder is that the encoding process does not require any explicit storage of the geometrical description of the cells. In exhaustic search algorithm, the search is performed sequentially on every code vector in the codebook, keeping track of the "best so far" and continuing until every code vector was tested. This method of pattern matching requires more time but the resulting compression is better than the nearest neighbour quantiser method.

2.2.4 Discrete Cosine Transform

Discrete Cosine Transform is widely used for many digital image compression techniques. Digital image compression technique developed by the Joint Photographic Experts Group (JPEG) is based on the DCT and predictive algorithm. The Consultative Committee for International Telegraph and Telephone (CCITT, now called International Telephone Union - ITU) and International Standard Organisation (ISO) formed this committee to develop compression standards for still and motion pictures [Wallace, 92].

In the JPEG algorithm there are mainly two image compression methods viz. predictive method which is carried out in the spatial (data) domain, and the transform method which is performed in the frequency domain [Leger, 91]. The predictive method is a lossless compression technique while the DCT-based method is a lossy compression technique. DCT-based method of compression is widely used as it is easier to implement and is suitable for a large number of applications including motion picture compression.

JPEG algorithm is applicable to a wide variety of applications. In most of the applications the JPEG algorithm is used as a library function. For example, NeXTstep is the standard operating environment on the NeXT computers and designed to support a wide range of applications. NeXTstep uses Tag Image File Format (TIFF). The JPEG algorithm is added to support TIFF file reading and writing facilities [Cockroft, 91]. JPEG is also used for Picture Archiving and Communication Systems (PACS) in medical imaging field [Kajiwara, 92]. A detail description of the JPEG standard is given in the next section.

2.3 JPEG Standard

There are three international standards available for image compression for different applications, viz. JPEG, MPEG, and P*64 [Quinnell, 93]. JPEG is intended for continuous-tone still images, MPEG is intended for video images, and P*64 is for video telephony. In this section the JPEG standard is described.

The first step in the JPEG algorithm is to locate data redundancy in the image pixel values. This is done by using the Discrete Cosine Transform, which is similar to the Fourier Transform but includes only the Cosine part of the function. Wavelet transforms can also be used with the JPEG standard in the video industry for on-line editing [Cornell, 93]. The Vector Quantisation (VQ) method is complicated by the need for code design. Therefore, coding with Vector Quantisation is slow as compared to coding with the JPEG algorithm. Vector Quantisation is more efficient when it is combined with other techniques. Therefore, the JPEG algorithm was chosen for our research purpose. The JPEG algorithm can be implemented in hardware as well as in software [Baran, 90].

Details of the DCT-based JPEG algorithm are given in section 2.3.1. The available hardware chips in VLSI implementation include the C-Cube Micro system's CL550 chipset, the SGS-Thomson's ST1140 CMOS chip [Leonard, 91], and Intel's Digital Video Interactive (DVI) chip [Vaaben, 91]. Section 2.3.2 describes the CL550 chip in brief.

JPEG has developed software implementation that can be used on general purpose machines and can be modified easily according to the application requirement. Section 2.3.3 describes the version-4 of modified JPEG software.

JPEG compressed file structure is described in section 2.3.4.

2.3.1 DCT-Based JPEG Algorithm

In the JPEG algorithm the source image is structured as follows [Wallace, 92]. A source image consists of 1 to 255 image components (depending upon resolution of the image). These components are sometimes called colours, spectral bands or channels. A colour image can be represented in many colour systems viz., Red Green Blue (RGB), YUV (Y for luminance or brightness, U and V for colour difference signals Y-R and Y-B respectively), Cyan Magenta Yellow and Black (CMYK) [Ang,

91]. Each component consists of a rectangular array of samples. A sample is defined to be an unsigned integer with the range $[0, 2^p - 1]$ or signed integer with the range $[-2^p, 2^p - 1]$, where p is sample precision in bits. The JPEG standard has defined the concept of a "data unit". A data unit is an 8 x 8 block of samples in DCT-based codecs. Generally, data units of image components are ordered from left-to-right and top-to-bottom.

Components of an image can be stored in one of two possible formats, namely interleaved format and non-interleaved format. If an image component is stored in the non-interleaved format, the data units are ordered in a pure raster scan sequence as shown in figure 2.2.

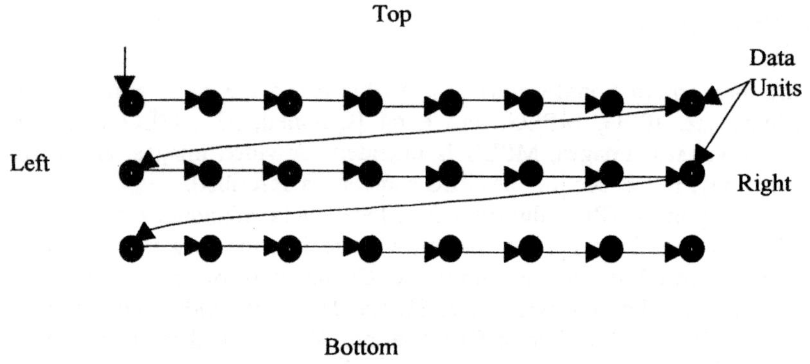

Figure 2.2 Non-interleaved data ordering

If the image has two or more colour components each one of these may be stored in an interleaved format. Each component Ci is partitioned into rectangular regions of Hi x Vi data units, as shown in the generalised example of figure 2.3. Regions within a component are ordered from left-to-right and top-to-bottom. Within a region also, data units are ordered from left-to-right and top-to-bottom.

2.3.1.1 DCT-based Compression Steps. Modified JPEG algorithm involves colour space conversion, Minimum Coded Unit (MCU) extraction, DCT, quantisation and encoding steps as shown in figure 2.4. Each of these step are described below.

1. Get the source file header information such as image format, image width, image height. Get the user specified parameters and generate quantisation table and Huffman encoding tables and initialise the JPEG output file header and marker.
2. Conversion from input image format to a standard internal format (either RGB or grayscale). Colour space conversion (eg. RGB to YCbCr). This is a null step for grayscale images.

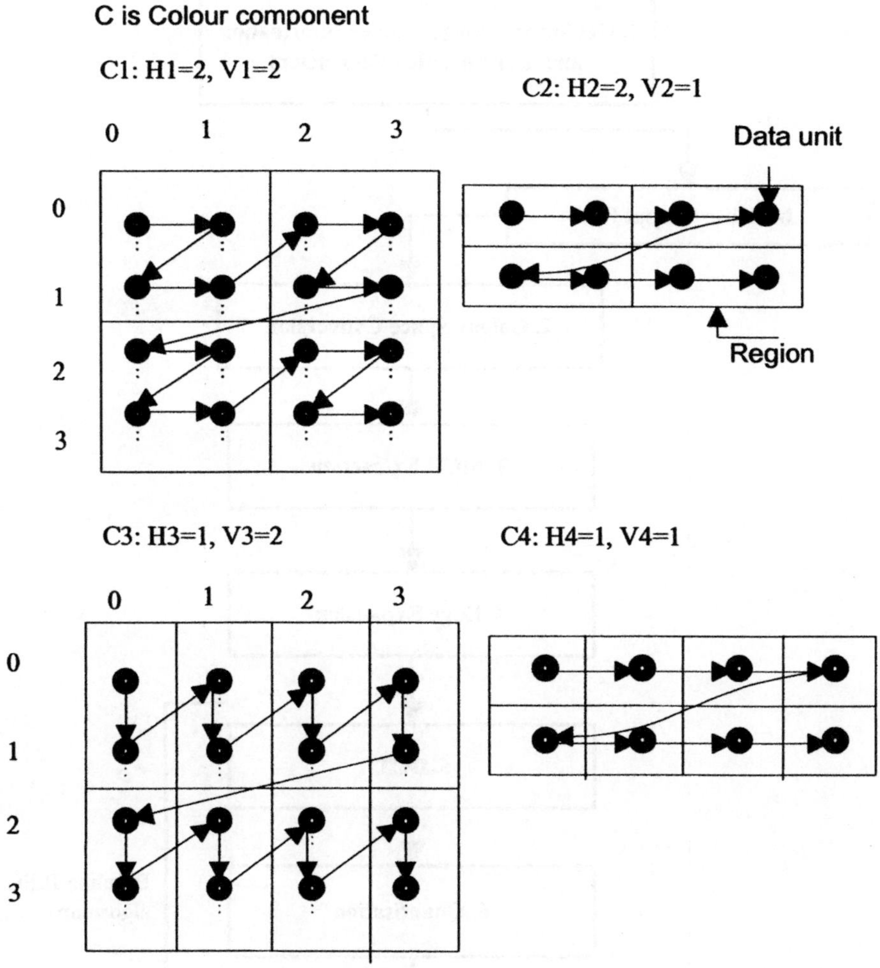

Figure 2.3 Interleaved image data ordering

The following steps (3 to 8) are performed once for each scan of a complete image, i.e. once if making a non-interleaved file, and more than once for an interleaved file.

3. Minimum Coded Unit (MCU) extraction, i.e. creation of a single sequence of 8 x 8 sample blocks.
4. Edge expansion. This step operates independently on each colour component.
5. DCT transformation of each 8 x 8 block.
6. Quantisation, scaling and zigzag reordering of the elements in each 8 x 8 block.
7. Huffman (or arithmetic) encoding of the transformed block sequence.
8. Output the JPEG file with required headers/markers.

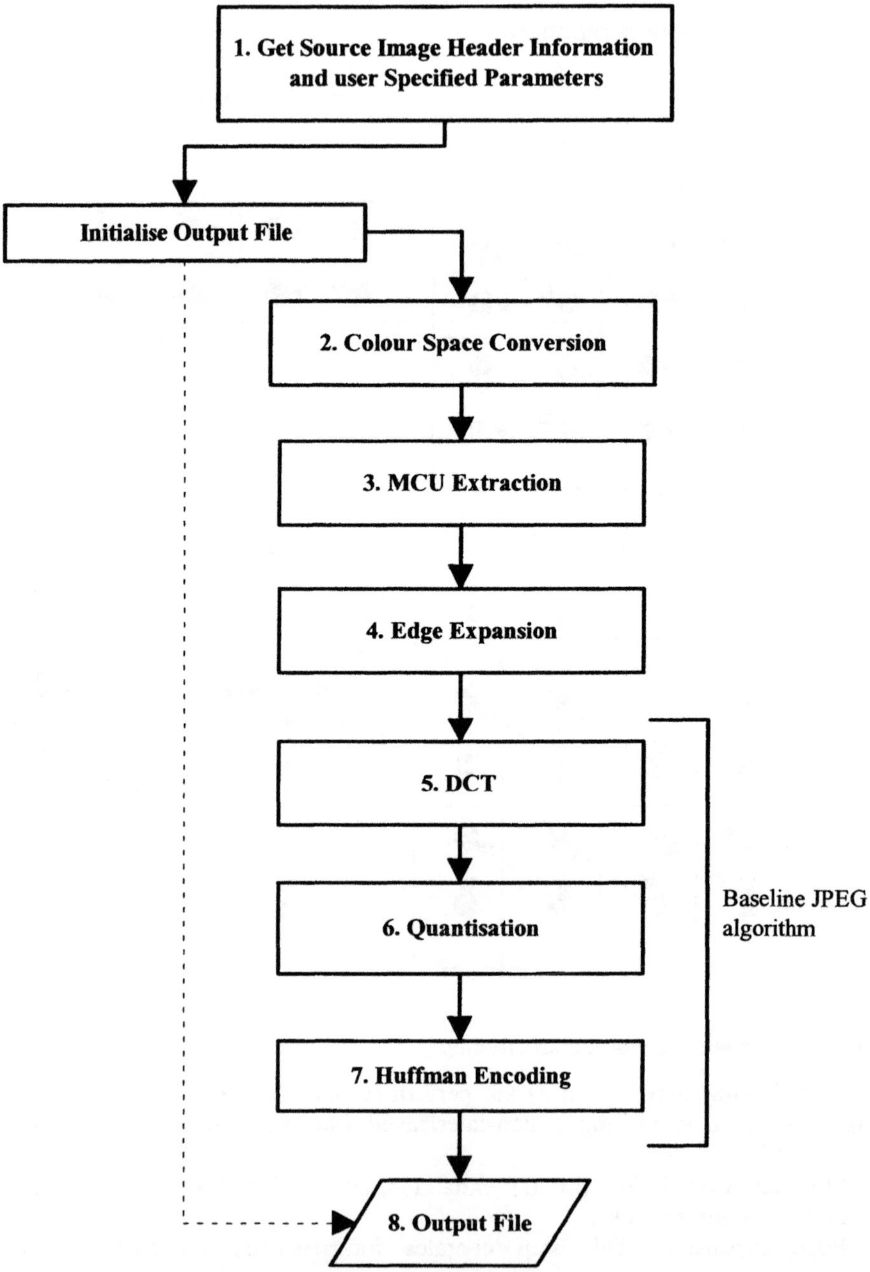

Figure 2.4 Modified JPEG algorithm

2.3.1.1a Input File and Parameters. The input image file to be compressed by the JPEG algorithm can be in either of the following formats: PPM (Pulse Pixel Map), GIF (Graphics Interchange Format), RLE (provided by Utah Raster Toolkit). Details of these formats differ but the type of information included in each of these is similar. In general, an input file contains information about the format used, image width, image height, maximum value of the sample, and the image data.

Parameters such as quality factor, smoothing factor, and sampling ratio are input by the user. For any unspecified parameters the default value is used by the software.

Quantisation tables are generated by taking into account the specified quality factor. AC and DC Huffman tables are also generated as part of the algorithm execution. Currently, the values generated are fixed by the JPEG standard. Though, it is possible to vary these tables to get compressed images of different compression ratios. The output file includes appropriate header along with the quantisation tables and Huffman tables. The subsequent step, colour space conversion, is explained in the next section.

2.3.1.1b Colour Space Conversion. The JPEG source image is divided into groups of rows. The number of rows in each group is equal to the maximum sampling factor. Each source image group (GrpSrcImg) is subjected to Colour Space Conversion (ClrSpcCnv) step. This step converts the input colour space of any format to the YCbCr format. The YCbCr format is defined by the CCIR 601-1 standard. For example, if the input image is in the RGB colour format, then the values of the Y, Cb, and Cr components can be calculated by the following formulae:

$$Y = 0.299900 * R + 0.58700 * G + 0.11400 * B$$
$$Cb = -0.016874 * R - 0.33126 * G + 0.50000 * B + MAXJSAMPLE/2$$
$$Cr = 0.50000 * R - 0.41869 * G - 0.08131 * B + MAXJSAMPLE/2.$$

Here MAXJSAMPLE is maximum value of sample in a source image. For example, in an 8-bit image, MAXJSAMPLE is 255.

2.3.1.1c MCU Extraction. Before applying the DCT function the input image is converted to a standard format and divided into 8 x 8 sample blocks.

The JPEG proposal defines the term "data unit" as a block of 8 x 8 samples and MCU to be the smallest group of interleaved data units. MCU extraction is used for better organisation of data units for interleaved image data. MCU is a group of data units taken from the same region of all image components.

A maximum of four components can be interleaved. And a maximum of ten data units are allowed in an MCU. Because of this restriction, not every combination of four components which can be represented in non-interleaved order within JPEG compressed image is allowed to be interleaved. The JPEG proposal allows some components to be interleaved and some to be non-interleaved within the same compressed image.

In the JPEG algorithm, the blocks of 8 x 8 samples and image components are processed sequentially; even though it is also possible to process image components simultaneously. In the non-interleaved format image components are independent of each other; whereas, in the interleaved format image components are dependent upon each other. In the interleaved format a maximum of four image components can be used in a single MCU. Thus, up to four image components can be processed at a time.

2.3.1.1d Edge Expansion. Edge expansion is used to make the number of samples in a block, a multiple of the MCU dimension. This is done by duplicating the right-most column and/or bottom-most row of pixels.

2.3.1.1e Discrete Cosine Transform (DCT). The Discrete Cosine Transform (DCT) is calculated for each block of the MCU. Output of the DCT F(u, v) gives orthogonal basis signals given by

$$F(u,v) = \frac{1}{4} C(u)\,C(v)\left[\sum_{x=0}^{7}\sum_{y=0}^{7} f(x,y) * \cos\frac{(2x+1)u\pi}{16} * \cos\frac{(2x+1)v\pi}{16}\right] \qquad (2.1)$$

where $C(u), C(v) = \begin{cases} =1/\sqrt{2} & u,v=0 \\ =1 & \text{otherwise,} \end{cases}$

and F(u,v) = Discrete cosine transformed signal.

Output of the DCT step for an 8 x 8 block of samples gives 64 coefficients. The zero frequency coefficient is called the DC coefficient and the remaining 63 coefficients are known as the AC coefficients. The DC coefficient is a measure of the average value of the 64 image samples.

2.3.1.1f Quantisation. Quantisation is achieved by dividing each DCT coefficient by its corresponding quantiser step size and rounding off to the nearest integer value, so that

Quantised value Q(u,v) = Integer Round (DCT coefficient / Quantiser step size). (2.2)

Quantiser step size is calculated with respect to the desired quality of the output image. This step is performed to achieve compression by representing DCT coefficients with no greater precision than is necessary to achieve a desired image quality. On performing quantisation, visually insignificant values are discarded.

2.3.1.1g Huffman Encoding. The first value to be encoded in a block is the DC coefficient. This DC coefficient is encoded as the difference between the DC term of this block and that of the previous block, in the encoding order shown in figure 2.5.

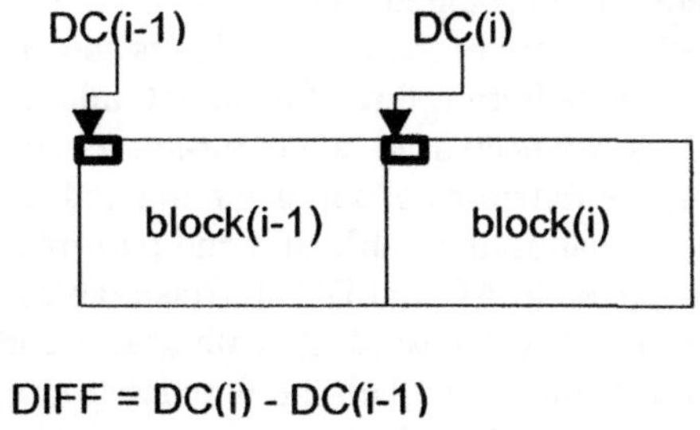

Figure 2.5 Encoding the DC coefficient

Coefficients other than the first one are called AC coefficients. These AC quantised coefficients are ordered in a "zig-zag" sequence, as shown in figure 2.6. Huffman encoding is done with a zig-zag ordering of the samples. This helps in placing the low frequency coefficients before the high frequency coefficients. Encoding requires one or more sets of Huffman code tables specified by the application or generated during the process of compression.

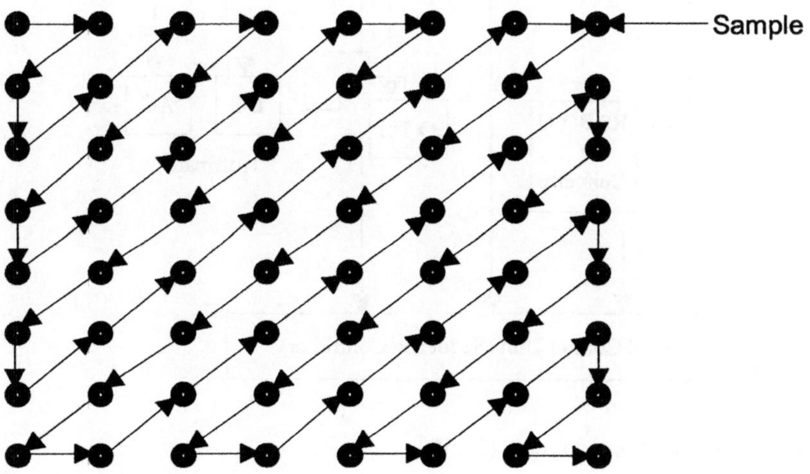

Figure 2.6 Zig-zag encoding order

2.3.1.1h JPEG Compressed File. The compressed image file contains a header, quantisation table, Huffman tables, and compressed data. The detailed structure of a compressed image file is described in section 2.3.4.

2.3.2 JPEG Hardware

The JPEG algorithm was implemented in hardware as a single VLSI chip [Ogawa, 92]. In this section the working and limitations of one such hardware implementation are briefly described.

C-Cube Microsystem has developed the CL550 chip for implementation of the JPEG algorithm. A block diagram of the CL550 chip is shown in figure 2.7. This chip consists of input buffer, output buffer, three functional-units and a sequence controller unit. Functional-unit-1 is connected to a colour-space converter (RGB to YUV converter) and performs the functions of sub-sampling and level shifting. Functional-unit-2 is connected to the quantisation table unit and performs DCT and quantisation. Functional-unit-3 is connected to AC and DC Huffman tables, and performs Huffman encoding and generates output compressed file with header and marker.

The sequence controller unit controls all the functional-units by giving instructions for execution in accordance with the values set by an internal register.

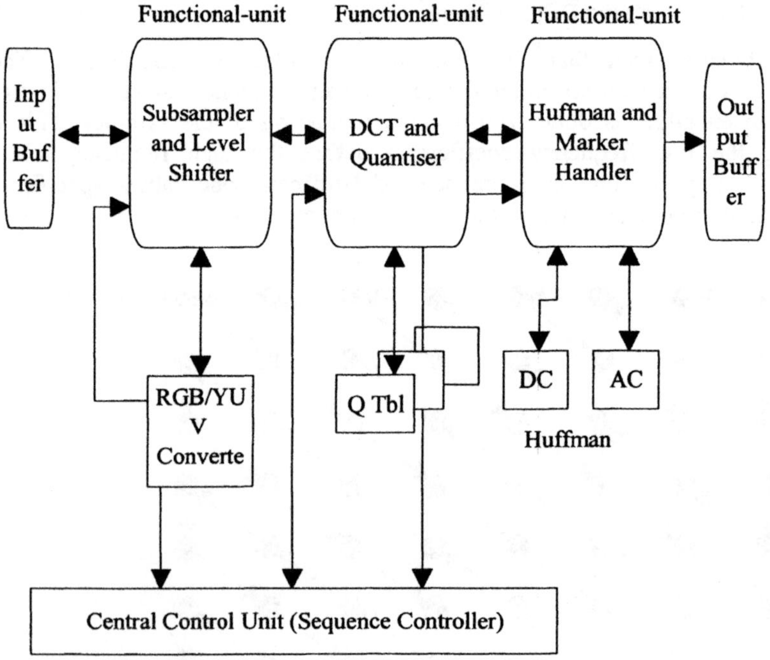

Figure 2.7 JPEG compressor chip (CL550)

Advantages of hardware implementation:

- Hardware implementation is faster than software based implementation.
- The chip design is tested extensively before fabrication.
- Operates at 30 MHz which is sufficient to compress and decompress real-time CCIR 601 format video [Ruetz, 93].

Disadvantages of hardware implementation:

- Output quality is limited, as only four tables are used in the chip.
- Huffman encoding tables are fixed. In software implementation it is possible to vary these tables to get compressed images of different compression ratios.
- The range of available subsampling ratios is limited, due to limited real-estate of the chip.

- Low flexibility of compression vis-a-vis quality and compression ratio.

Many of the limitations of hardware implementation can be overcome in software based implementations. Main advantages of software based implementation are:

- Flexibility, for variation of parameters such as quality of output image.
- User specified quantisation table and sampling ratio etc.
- Can be implemented on general purpose machines.
- Can be implemented on a variety of parallel computer architectures.

2.3.3 DCT-based JPEG Software

JPEG software packages are available from organisations such as C-Cube Microsystems and Kodak. JPEG software is available in many versions. The VER-4 of the software was used. The VER-4 software includes routines for down-sampling in addition to the standard JPEG steps shown in figure 2.8.

Execution of the JPEG compression software is initiated by the user command "CJPEG". This command is followed by a list of parameters such as quality factor, sampling factor, smoothing factor, input file name and output file name. The software generates quantisation table form the user specified quality factor and stores it in a buffer. This table is used in the quantisation step, and is also stored in the output file.

The JPEG software takes information such as image format, image-width, image-height and maximum value of the sample from the input file header. Number of data units and the number of MCUs is calculated from the image-width and image-height. Then the software reads a group of image rows equal to the maximum horizontal sampling factor. Each group is subjected to colour space conversion and downsampling steps. The colour space conversion process depends upon the image format.

Smoothing operation is performed on the image to clean up a dithered input file. The smoothing factor (SF) determines the level of smoothing performed. User can specify the smoothing factor ranging from 1 to 100. Each input pixel P is replaced by P' a weighted sum of itself and its eight neighbours as given in the following formula,

$$P' = ((\text{Sum of eight neighbouring samples}) * SF + P(1-8*SF)) / 9$$

where $SF = Sf / 1024$

Sf = User specified Smoothing factor.

In the next step smoothed data units are subjected to down-sampling. Down-sampling is used to reduce the total number of samples. For example, a 2 x 2 array of blocks may be reduced to a single block by replacing groups of four samples by their average value.

Down-sampled data units are subjected to edge expansion. MCUs are extracted from the edge expanded rows. DCT and quantisation is performed on each data unit of the MCU. Quantised MCU is then subjected to a Huffman encoding step. The encoded image data is stored initially in a buffer and then in an output file.

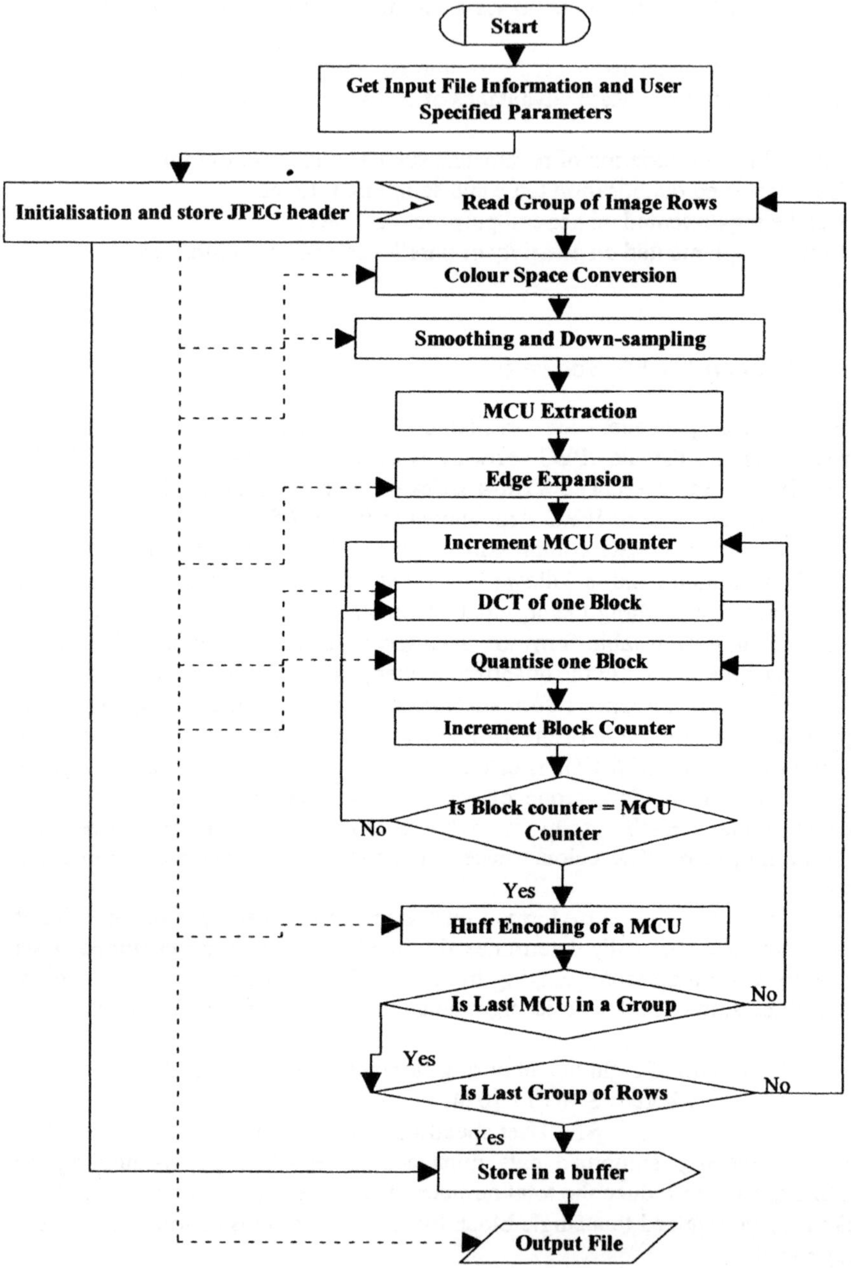

Figure 2.8 JPEG software routines

2.3.4 Compressed JPEG Data Structure

This section describes the JPEG compressed data file structure. JPEG algorithm consists of non-interleaved and interleaved data ordering as described in section 2.3.1. In this section the non-interleaved data ordering is considered for grey scale images. JPEG compressed data structure for non-interleaved, greyscale image is as follows [Pennabaker, 93].

Each compressed data file consists of a header, followed by the compressed image data, and terminated by an End of Image (EOI) marker, as shown in table 2.1.

Table 2.1 JPEG compressed data file structure.

JHI	JPEG Header Information
Compressed Data Blocks	
EOI	End Of Image Marker

The JPEG header contains four groups of information: quantisation table, frame header, Huffman table, and scan header. This JPEG header is followed by blocks of compressed image data, and EOI as shown in table 2.2.

Table 2.2 JPEG compressed data structure for non-interleaved greyscale

SOI	SOI-Start OF Image marker
DQT, length, quantisation table definition(s)	DQT- Digital Quantisation Table
SOFn, length, frame parameters	SOF-Start Of Frame
DHT, length, Huffman table definition	DHT-Digital Huffman Table
SOS, length, scan parameters	SOS-Start Of Scan
Compressed Block1 data	
Compressed Block2 Data	
Compressed Block3 Data	
Compressed Block4 Data	
.....etc.....	
Compressed Blockn Data	
EOI	EOI-End Of Image marker

A more detailed description of each component of the compressed file is given in the following section.

2.3.4.1 Quantisation Table Specification. The quantisation table is included in the compressed data file. This table determines the quality of the output image and is determined by the parameters following the **DQT** marker shown in table 2.3. In this structure **Pq** specifies the precision of the quantisation table elements (0 for 8-bit precision, 1 for 16-bit precision). **Tq** is the quantisation table identifier and may have values from 0 to 3. **Qk** is the quantisation table elements. It may have values from 1 to the maximum value permitted for the specified precision [Pennebaker, 93].

Table 2.3 Quantisation data segment structure in the JPEG algorithm:

Parameter	Symbol	Size (bits)
Marker (X'FFDE')	DQT	16
Quantisation table definition length	Lq	16
For each quantisation table:		
Quantisation table element precision	Pq	4
Quantisation table identifier	Tq	4
Quantisation table element (k=0, 1, 2, ..., 63)	Qk	8 or 16

2.3.4.2 Huffman Table Specification. The **DHT** marker segment provides a mechanism for specifying the Huffman code tables as shown in table 2.4. In this structure, the table class **Tc**, is 0 for DC and lossless coding, whereas **Tc** is 1 for AC code tables. **Th** is the code table identifier, it may have values from 0 to 3. The Huffman table is specified by identifying the number of codes of each length from 1 to 16. Huffman code of each length is specified by **Vij**.

Table 2.4 Huffman data structure

Parameter	Symbol	Size (bits)
Marker (X'FFC4')	DHT	16
Huffman table definition length	Lh	16
For each Huffman table:		
Table Class	Tc	4
Huffman table identifier	Th	4
Number of Huffman codes of length i for i = 1, ..., 16	Li	8
Value associated with each Huffman code for i = 1, ..., 16; j = 1, ..., Li	Vij	8

2.3.4.3 Frame Header. The JPEG algorithm divides the image into a number of frames and scans. For non-interleaved data ordering, a single frame is used, whereas in interleaved data ordering, a sequence of frames are used for encoding. The mode of encoding is represented by frame header structure in compressed data structure as shown in table 2.5.

The frame header length **Lf** gives the length in bytes of the frame parameters. Sample precision **P** gives the precision of sample in bits. Available values for sample precision are 8-bit, 12-bit or 16-bit. Number of lines **Y** represents the number of raster lines after the edge expansion step. **X** specifies the number of samples per raster line in a frame. **Nf** specifies the number of components in a single frame; it can range from 1 to 255. Each frame can be specified by its component identifier, horizontal and vertical sampling factor for MCU extraction, and quantisation table used for encoding the component.

Table 2.5 Frame header data structure

Parameter	Symbol	Size (bits)
Marker (X'FFCO-3, 5-7, 9-B, D-F')	SOFn	16
Frame header length	Lf	16
Sample precision	P	8
Number of lines	Y	16
Number of samples/line	X	16
Number of components in frame	Nf	8
Frame component specification (i=1, ..., Nf)		
Component identifier	Ci	8
Horizontal sampling factor	Hi	4
Vertical sampling factor	Vi	4
Quantisation table destination selector	Tqi	8

2.3.4.4 Scan Header. Each frame may have a number of scans, and the scan header information is included in a compressed data structure as shown in table 2.6. The number of components in a scan depends upon the mode of image scanning (non-interleaved and interleaved). In non-interleaved mode of scanning a scan has only one data unit, and in interleaved mode of scanning a scan has more than one data units.

Table 2.6 Scan header data structure

Parameter	**Symbol**	**Size** (bits)
Marker (X'FFDA')	SOS	16
Scan header length	Ls	16
Number of components in scan	Ns	8
Scan component specification(k=1,...,Ns)		
Scan component selector	Csk	8
DC entropy coding table selector	Tdk	4
AC entropy coding table selector	Tak	4
Start of spectral selection or predictor selection	Ss	8
End of spectral selection	Se	8
Successive approximation bit position high	Ah	4
Successive approximation bit position low or point transform	Al	4

Ls gives the length of scan header in bytes. **Ns** specifies the number of components in a scan, it ranges from 1 to 4. Each component in a scan is specified by scan component selector **Csk** and entropy coding table selectors **Tdk** and **Tak**. Scanning takes each data unit from the band of MCU coefficients in zig-zag order. The start of scan selection **Ss** identifies starting index of this band. **Se,** the end of scan selection identifies the index of the last coefficient in the spectral band.

2.4 Block Comparator Enhancement to the JPEG Algorithm

In the JPEG algorithm all image blocks are processed individually. These blocks of compressed image are stored sequentially as shown in figure 2.9.

The JPEG algorithm divides the input image into a number of blocks. These blocks are arranged in i rows and j columns. These blocks are processed sequentially from Block-1 to Block-n from left to right and top to bottom [Papathanassiadis, 92]. Compressed data blocks are stored sequentially, as shown in the compressed data file structure of figure 2.9. The compressed image file begins with a header, and an End of Image (EOI) marker is placed at the end of this file.

In many types of images, there is the possibility of having one or more similar blocks in the image. Improvement in performance of the compression algorithm can be achieved by locating similar blocks in the image. The Block Comparator Technique is proposed to enhance the performance of the JPEG algorithm. An overview of the Block Comparator Technique is given in this section.

The logic of the Block Comparator Technique is shown in figure 2.10. The input file is divided into the required number of blocks, and each block is identified by a block number. Then all blocks are passed through the block comparator.

The block comparator algorithm is shown as a flow chart figure 2.11. Structure of the compressed data file is shown in figure 2.12. Similar blocks are identified by maintaining a Match List and a Reference List (figure 2.12). Each unique block is given a Unique Block Number. This Unique Block Number is stored in the Reference List along with the (original) Block Number. The Match List matches each non-Unique Block Number to the Number of the Unique Block that is similar.

In the block comparator step, first block is taken as Unique Block Number 1. Each block is compared to the existing unique blocks. If there is a match found, the Block Number is store in the Match List followed by the Unique Block Number. If there is no match found, that block is identified with a new Unique Block Number and stored in the Reference List. This process is repeated for all blocks. The Match List matches each block to similar unique blocks. Reference List is the list of all unique blocks. Compression is performed only on the unique blocks, and compressed image blocks are stored in the Compressed Image File after the Reference List as shown in figure 2.12. Image header and EOI marker are placed at the beginning and the end of the file respectively.

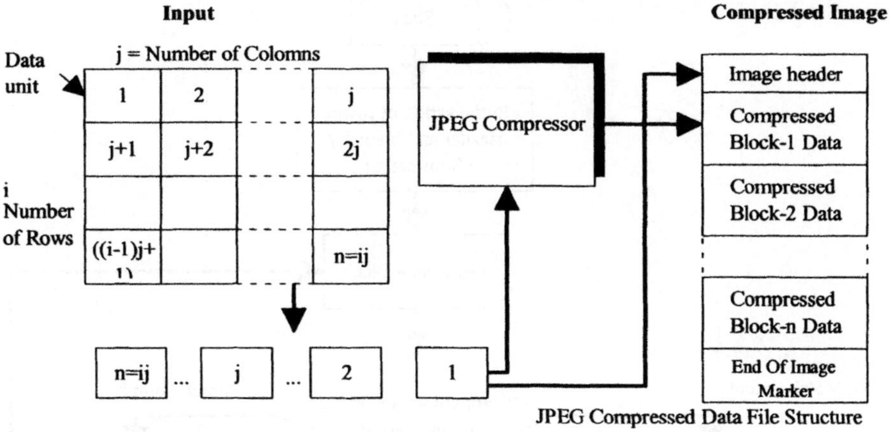

Figure 2.9 Sequential processing and storage of image blocks in the JPEG compression standard

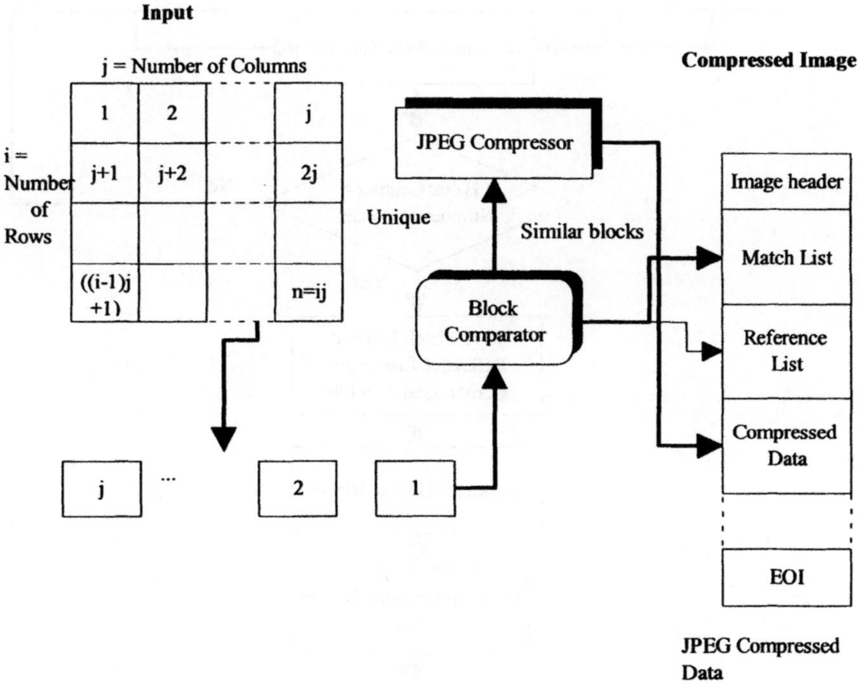

Figure 2.10 Block comparator enhancement to the JPEG compression algorithm

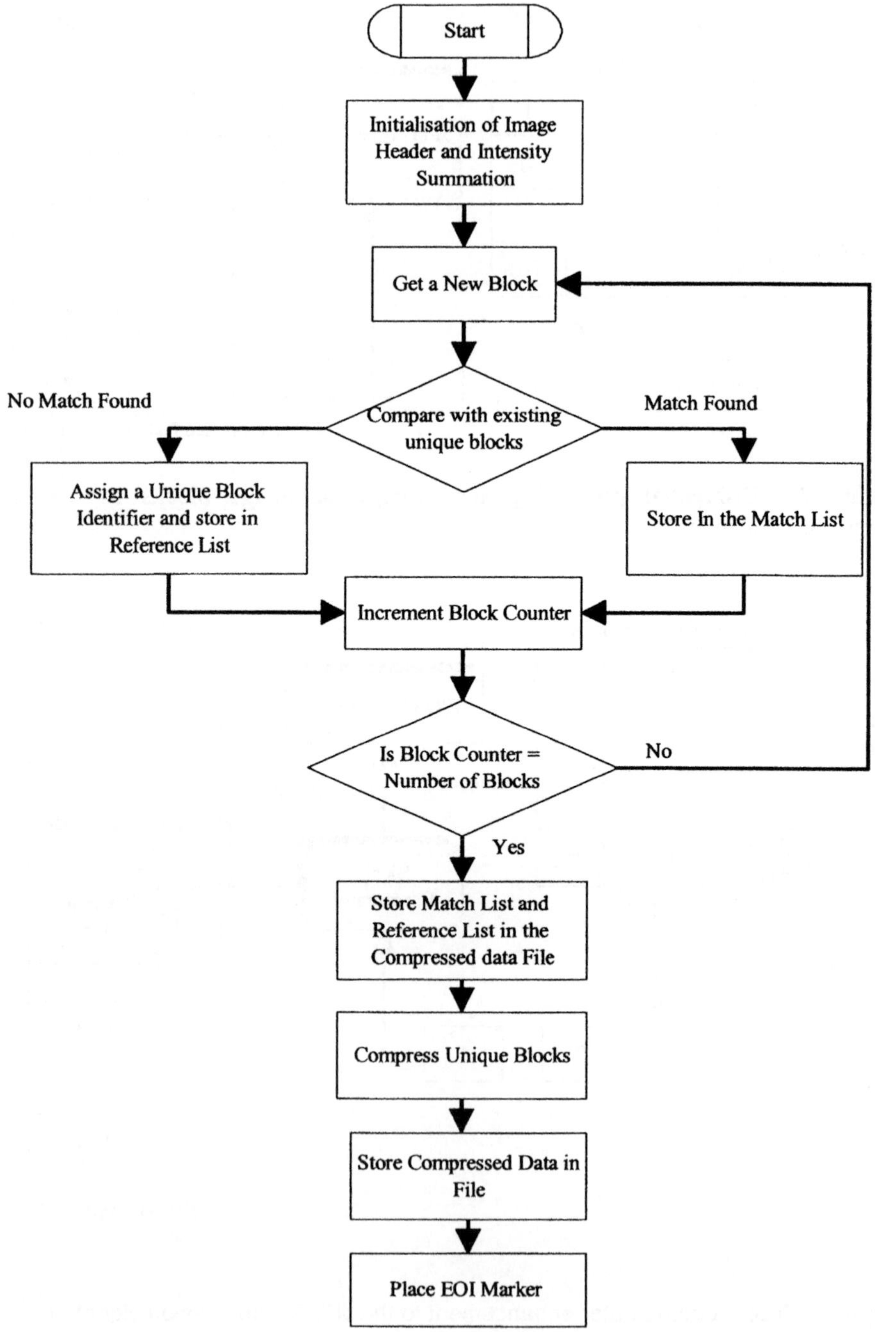

Figure 2.11 Flow chart of Block Comparator Technique

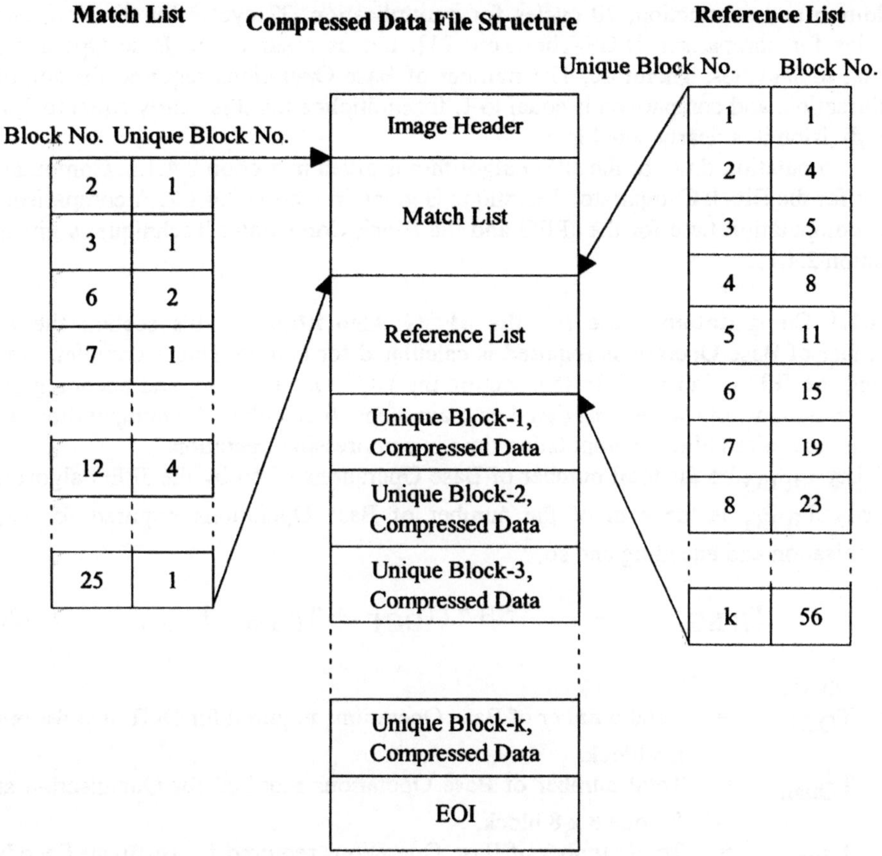

Figure 2.12 Compressed data file structure in Block Comparator Technique

The Block Comparator Technique improves the speed of the compression and reduces the size of the compressed data file. These two factors are discussed in the following sections.

2.4.1 Comparison of the JPEG Algorithm and Block Comparator Technique Execution Times

In this section the execution time of the JPEG algorithm is calculated with and without the proposed Block Comparator Technique. The computation time is calculated in terms of number of the arithmetic operations such as additions, subtractions, multiplications, divisions, and comparisons. Each arithmetic operation is equated to a number of Base Operations. The number of equivalent Base Operations for each arithmetic operation can be determined for specific processors. Transputer IMS T805 processor is chosen for calculating the number of Base Operations. For 64-

bit floating point operation, Transputer IMS T805 processor takes 7 cycles for addition and subtraction, 20 cycles for multiplication, 32 cycles for division, and 7 cycles for comparison [SGS-Thomson, 91]. Let us assume one Base Operation is equal to 7 cycles. Therefore, The number of Base Operations required for addition, subtraction, and comparison is equal to 1, for multiplication it is nearly equal to 3, and for division it is nearly equal to 5.

Computation time for the JPEG algorithm is given in section 2.4.1.1. Computation time for the Block Comparator Technique is given in section 2.4.1.2. A comparison of the computation time for the JPEG and the Block Comparator Technique is given in section 2.4.1.3.

2.4.1.1 Computation Time for the JPEG Algorithm. In this section the total number of Base Operations required is calculated for compressing a complete image using the JPEG algorithm. In this section the DCT, quantisation, and encoding steps of JPEG algorithm were considered as main steps to calculate the computation time, as these required higher computation time in compression operation.

Let T_{JPEG} be the total number of Base Operations taken by the JPEG algorithm. Then T_{JPEG} is the sum of the number of Base Operations required for DCT, quantisation and encoding and so,

$$T_{JPEG} = NB * (T_{DCT} + T_{Quan} + T_{Enco}) \tag{2.3}$$

where,

T_{DCT} = Total number of Base Operations required for DCT step for one 8 x 8 block,

T_{Quan} = Total number of Base Operations required for Quantisation step for one 8 x 8 block,

T_{Enco} = Total number of Base Operations required for Huffman Encoding for one block,

NB = Number of Blocks.

Discrete Cosine Transform step: DCT step consists of two operations. First operation is subtraction of each sample from half of the maximum sample value. The second operation is the DCT function itself. Therefore, the total number of Base Operations required for the DCT step can be calculated as

$$T_{DCT} = T_{CS} + T_{dct} \tag{2.4}$$

where,

T_{CS} = Total number of Base Operations required for subtraction operation for one
8 x 8 block
= 64 Subtractions
= 64 Base Operations,

T_{dct} = Total number of Base Operations required for DCT function for one 8 x 8 block.

The Independent JPEG Group (IJG) source code uses 12 multiplications and 32 additions to perform DCT function for a one dimension (1-D) DCT. A 2-D DCT in IJG software is done by performing 1-D DCT on each row followed by 1-D DCT on each column. Therefore, we need 16 1-D DCTs to perform 2-D 8 x 8 DCT as given in equation 2.5,

$$\begin{aligned} T_{dct} &= 16\,(12 \text{ multiplications} + 32 \text{ additions}) \\ &= 192 \text{ multiplications} + 512 \text{ additions} \\ &\cong 3 * 192 + 512 \text{ Base Operations} \\ &\cong 1088 \text{ Base Operations.} \end{aligned} \tag{2.5}$$

By substituting the value of T_{CS} and T_{dct} into equation 2.4 we get the total number of Base Operations for DCT step as 1152 Base Operations, as

$$\begin{aligned} T_{DCT} &\cong 64 + 1088 \\ &\cong 1152 \text{ Base Operations.} \end{aligned} \tag{2.6}$$

Quantisation: Quantisation step involves the division of each sample by its quantiser step. An 8 x 8 block, quantisation takes 64 division, thus,

$$\begin{aligned} T_{Quan} &= 64 \text{ Divisions} \\ &\cong 64 * 5 \text{ Base Operations} \\ &\cong 320 \text{ Base Operations.} \end{aligned} \tag{2.7}$$

Huffman encoding: For the DCT and the quantisation steps the required number of Base Operations could easily specify. Huffman encoding requires many arithmetic and logic operations such as multiplication, increments, shift operations, subtractions. Therefore it is difficult to specify the exact number of arithmetic operations for the Huffman encoding step. The experimental results showed that the total time required for Huffman encoding is about 60% of the compression time. The remaining 40% of the compression time is required for DCT and quantisation. From these experimental results, the number of Base Operations required are calculated for Huffman encoding step as,

$$\begin{aligned} T_{Enco} &= 60 * \left(\frac{T_{DCT}+T_{Quan}}{40}\right) \\ &= 60 * \left(\frac{1152+320}{40}\right) \\ &\cong 2208 \text{BaseOperations} \end{aligned} \tag{2.8}$$

By combining equations 2.6, 2.7 and 2.8 relating the total number of base operations for one block, the total number of Base Operations required for the JPEG algorithm can be calculated as,

$$T_{JPEG} = NB * (T_{DCT} + T_{Quan} + T_{Enco}) \tag{2.9}$$

$$= NB * [(T_{dct} + T_{cs}) + T_{Quan} + T_{Enco}$$

$$= NB * [1152 + 320 + 2208]$$

$$\cong NB * [3680] \text{ Base Operations.}$$

The JPEG algorithm takes 3680 Base Operations to perform compression operation on one block. The complete image consists of NB number of blocks. Therefore NB x 3680 Base Operations are required for compressing a complete image.

2.4.1.2 Computation Time Taken for Block Comparator Algorithm. In this section the total number of arithmetic operations required are calculated for compressing a complete image using the Block Comparator Technique. The number of arithmetic operations required for the Block Comparator algorithm is the sum of the number of arithmetic operations required for Comparison and the number of arithmetic operations required for Compression, and so,

$$T_{BCT} = T_{BC} + T_{JPEG} \tag{2.10}$$

where,

T_{BC} = Total number of Base Operations required for (Block) Comparison,

T_{JPEG} = Total number of Base Operations required for Compression.

The number of Base Operations required for the Compression step T_{JPEG} can be calculated from equation 2.3 by replacing the Number of Blocks (NB) by the Number of Unique Blocks (NUB).

The number of Base Operations required for the Block Comparator step are calculated in this section. Block Comparison step consists of following three main steps,

1. Summation of sample intensity value in each block
2. Block intensity comparison
3. Sample-by-sample comparison of blocks

The additional steps required in the Block Comparator Technique are shown in figure 2.13. The number of Base Operations required for Block Comparison, T_{BC} is given by,

$$T_{BC} = T_{sum} + T_{intcomp} + T_{sampcomp} \tag{2.11}$$

where,

T_{BC} = Total number of Base Operations for Block Comparison,

T_{sum} = Total number of Base Operations for summation of samples in all image blocks,

$T_{intcomp}$ = Total number of Base Operations for block intensity value comparison,

$T_{sampcomp}$ = Total number of Base Operations for comparing samples of a block with those of existing Unique Blocks.

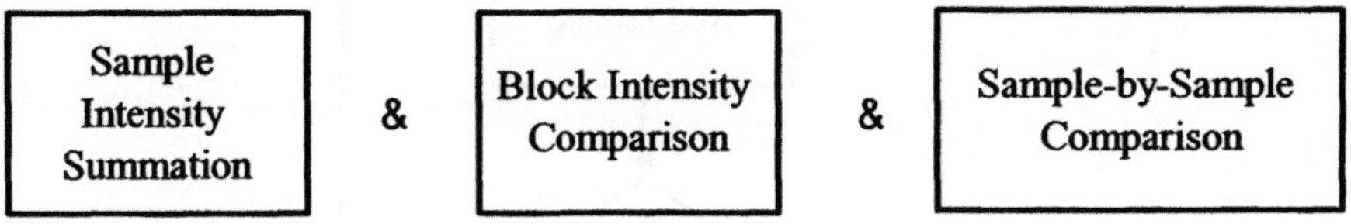

Figure 2.13 Additional steps required in the Block Comparator Technique

Sample intensity summation: Algorithm for summation of samples is shown in the flow chart of figure 2.14. In the summation step, the first block is called Block-1. Sample values of this block are summed sample-by-sample, and the summation result is stored in a Block Intensity List (BIL) as the Block Intensity Value-1 (BIV-1). This process continues for all blocks. Block Intensity List consists of Block Intensity Values for all blocks.

The total number of arithmetic operations required for summation process, T_{sum} is given by,

$$\begin{aligned} T_{sum} &= NB * T_{sampsum} \\ &= NB * 64add \\ &= NB * 64 \text{ Base Operations} \end{aligned} \tag{2.12}$$

where,

$T_{sampsum}$ = Total number of Base Operations required for the summation of sample values in any image block.

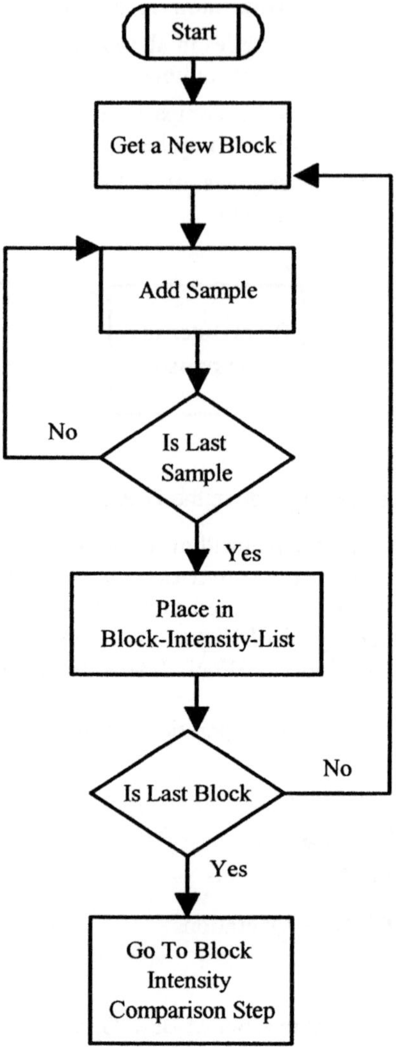

Figure 2.14 Flow chart for summation step

Block intensity comparison: In the Block Intensity Comparison step each Block Intensity Value (BIV) is taken from the Block Intensity List. This BIV is compared to each of the existing Unique Block Intensity Value (UBIV). If the BIV is equal to any of the UBIVs, then this BIV is stored in an Equal Intensity List (EIL). If this BIV is not equal to any of the UBIVs for the unique blocks stored in the Reference List then this block number is stored in the Reference List and the BIV is stored in the UBIV List. The Flowchart for Block Intensity Comparison is shown in figure 2.15.

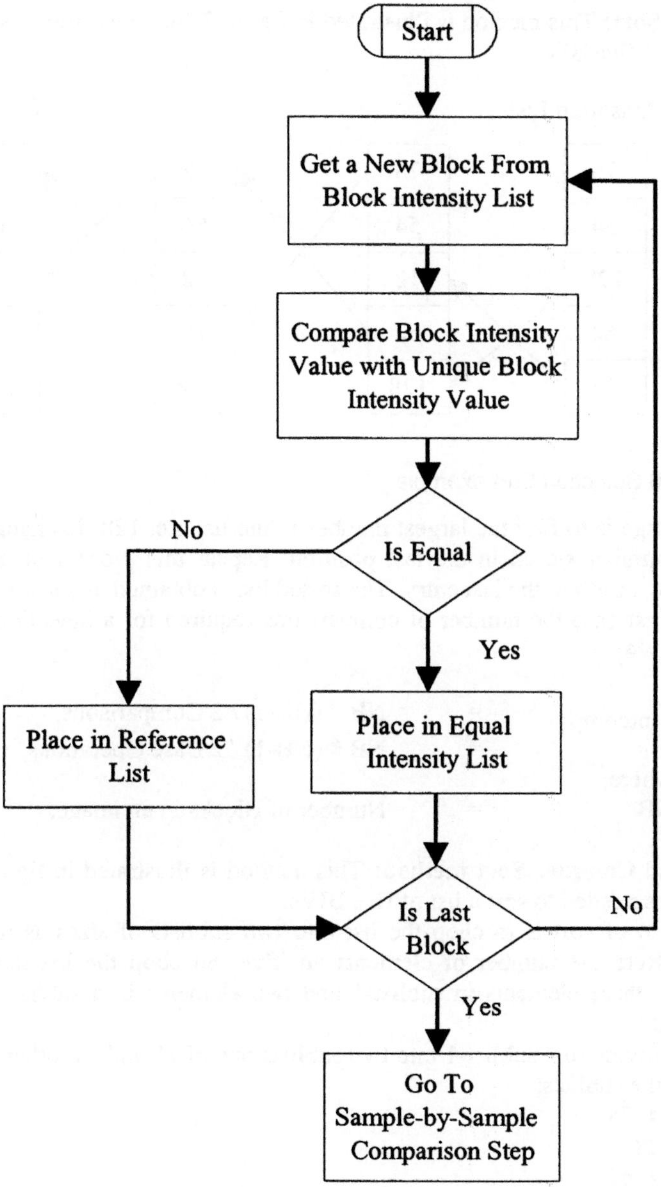

Figure 2.15 Flowchart for block intensity comparison step

Block comparison can be done either on a list of sorted values or on a list of unsorted values. The sorted list is used for comparison because it helps in grouping equal intensity values. There are many sorting techniques. The Selection Sort and the Divide and Conquer sort methods were chosen in this chapter [Kruse, 94]. These two sort methods are explained in the following sections.

Selection Sort: This method is illustrated in figure 2.16, which shows steps needed to sort a list of five BIVs.

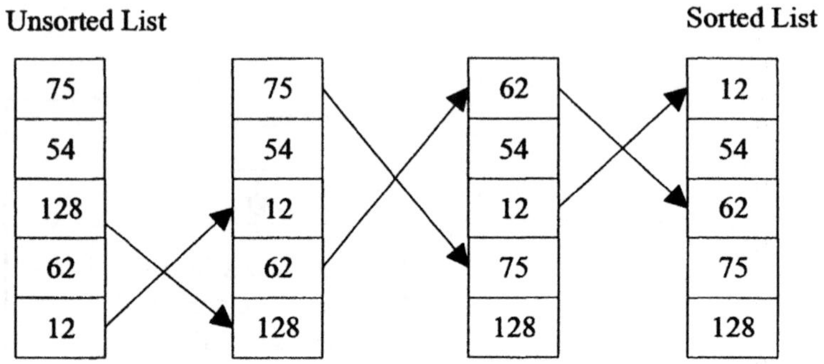

Figure 2.16 Selection Sort example

First stage is to find the largest number in the list, i.e. 128. Exchange this number with the number stored in the last position. Repeat this process on the shorter list obtained by omitting the last entry. The sorted list is obtained as shown in the last list.

For worst case the number of comparisons required for a Selection Sort is given by [Kruse, 94],

$$T_{intcomp} = NB * (NB-1) / 2 \text{ Comparisons,} \quad (2.13)$$
$$= NB * (NB-1) / 2 \text{ Base Operations,}$$

where,

NB = Number of Blocks in an image.

Divide and Conquer Sort method: This method is illustrated in figure 2.17, which shows steps needed to sort a list of five BIVs.

The step of sort is to chop the list into two sublists of sizes as nearly equal as possible. Here the number of elements are five. So chop the list into two sublists consists of three elements in sublist-1 and two elements in a sublist-2 as shown in figure 2.18.

Again divide first sublist-1 into two sublists one of 75 and 54, other sublist of 128. Then sort the sublists:

54, 75
128
12, 62

Then merge the sorted sublists. First merge the sublist-1 and sublist-2 and sort it.

54, 75, 128 ---------> 54, 75, 128
Merge this list with sublist-3.
54, 75, 128 and 12, 68 ------------> 12, 54, 68, 75, 128

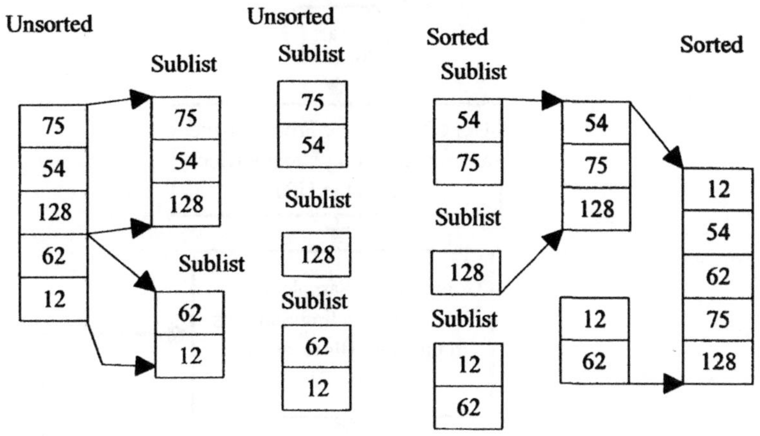

Figure 2.17 Divide and Conquer Sort example

The worst case number of comparisons required for Divide and Conquer sort method is given by [Kruse, 94],

$$T_{intcomp} = NB \lg NB \text{ Comparisons} \quad (2.14)$$
$$= NB \lg NB \text{ Base Operations,}$$

where,

NB = Number of Blocks in an image.

Sample-by-Sample comparison: Algorithm for sample-by-sample comparison of two blocks is shown in the flow chart of figure 2.18. A Block is picked from the Equal Intensity List (EIL) and compared with the Unique Block sample-by-sample. The two blocks are said to 'match' if and only if all samples in one block have exactly the same values as the corresponding samples in the other block. If the block being compared matches with any of the unique blocks identified, it is placed in the Match List, else it is entered in the Reference List. This process continues for all the blocks.

For some type of images closely matching blocks may be acceptable. But, in the current investigation it is not looked into this possibility.

The total number of arithmetic operations required for the sample comparison step is given by,

$$T_{sampcomp} = N_{BEIL} * (T_{sampblock}) \quad (2.15)$$

where,

N_{BEIL} = Number of blocks in Equal Intensity List (EQL),

$T_{sampblock}$ = Total number of Base Operations required for sample-by-sample comparison of one block ,

= 64 comparisons (Maximum),

≅ 64 Base Operations (Maximum).

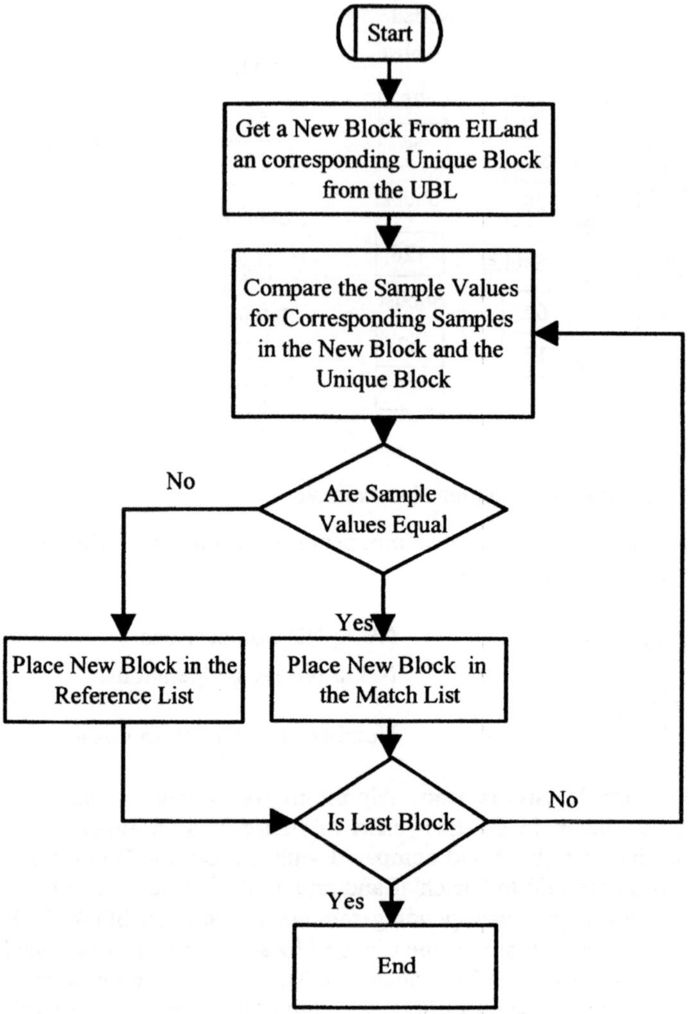

Figure 2.18 Flowchart for Sample-by-Sample Comparison step

Total number of base operations for the Block Comparator Technique:

By substituting equation 2.11 in equation 2.10, we get

$$T_{BCT} = T_{sum} + T_{intcomp} + T_{sampcomp} + T_{JPEG} \qquad (2.16)$$

By substituting the equation (2.12), (2.13) and (2.15) in equation (2.16), we get the equation (2.17) for T_{BCT}, using Selection Sort method,

$$T_{BCT} = NB * 64 + (NB*(NB-1))/2 + NBEIL * 64 + NUB * 3680 \quad (2.17)$$

By substituting the equation (2.12), (2.14) and (2.15) in equation (2.16), we get the equation (2.18) for T_{BCT} using Divide and Conquer sort method.

$$T_{BCT} = NB * 64 + (NB * lgNB) + NBEIL * 64 + NUB * 3680 \quad (2.18)$$

2.4.1.3 Comparison of Computation Time for the Non-Block Comparator Technique and the Block Comparator Technique. The JPEG algorithm is called as Non-Block Comparator Technique (NBCT) in this book. The Block Comparator Technique (BCT) using Selection Sort method for intensity comparison is called as Selection Sort method and BCT using Divide and Conquer Sort method is called as Divide and Conquer Sort method in this section.

The speed improvement obtained by the Block Comparator Technique over the Non-Block Comparator Technique can be represented by a factor called the Speed Improvement Factor (SIF). SIF is defined as the ratio of total number of Base Operations required for the Non-Block Comparator Technique (NBCT) to the total number of Base Operations required for the Block Comparator Technique (BCT), as given by

$$SIF = \frac{T_{NBCT}}{T_{BCT}} . \quad (2.19)$$

Values of SIF using the Selection Sort method are given in table 2.7. For the Divide and Conquer Sort method SIF values are given in table 2.8. SIF values obtained by using the direct Sample-by-Sample comparison method are given in table 2.9. In each of these tables the SIF values are calculated for three image sizes, viz. NB = 256, 4266, and 15594. For each image size SIF values are calculated for Number of Similar Blocks (NSB) in the image equal to 0%, 10%, 30%, 50%, 75%, 85%, 90%, 95% and 100%. Graph of SIF versus NSB using the Selection Sort method, is shown in figure 2.19. The graph for the Divide and Conquer Sort method is shown in figure 2.20, and graph for the block comparison using direct Sample-by-Sample Comparison method is shown in figure 2.21.

Conclusions are derived by comparing SIF values obtained for the three different methods, and also by comparing SIF values for each method individually. These conclusions are given in the following sections.

Common conclusions for all methods: The Speed Improvement Factor (SIF) is less than one for a zero number of similar blocks irrespective of the method used and the image size. This result is expected, as there is no speed improvement when there are no similar blocks; because there is additional computation time required for the block comparison step. Therefore, the Block Comparator Technique will add unwanted computational overhead if there are no similar blocks in an image.

SIF values for all image sizes increase monotonically with increase in the value of NSB. For most of the methods studied SIF is greater than one for NSB >= 10%. This indicates that the Block Comparator Technique delivers dividends even for a small number of similar blocks. The maximum SIF was obtained for 100%, that indicates image is a blank paper.

Conclusion for the BCT using the Selection Sort method: SIF values for NB = 256 is greater than one for NSB in the range of 10% to 100%. Whereas, for NB = 4266 SIF is less than one even upto 50% similar blocks. On the other hand for NB = 15594 SIF is less than one for all values of NSB. This indicates that there is no benefit in using the Block Comparator Technique in conjuction with the Selection Sort method for large images. Selection Sort method is suitable only for small image size.

Conclusion for the BCT using the Divide and Conquer Sort method: SIF is greater than one for NSB >= 10% and increases monotonically. SIF is almost equal for all image sizes for the same values of NSB. Therefore, the Divide and Conquer method is suitable for all image sizes.

Conclusion for the BCT using the direct Sample-by-Sample Comparison method: The SIF values for all image sizes and values of NSB are almost equal to the SIF values for the Divide and Conquer method, except for NSB=100. For NSB=100, SIF is greater for the Sample-by-Sample comparison method than that for the Divide and Conquer Sort method.

Divide and Conquer Sort versus Sample-by-Sample Comparison: Block comparison using Divide and Conquer Sort method is used to sort the image blocks according to the intensities of these blocks. This helps to group blocks with equal intensities. Equal intensity blocks can then be distributed on a parallel computer so as to balance the work load on all processors. This increases the improvement in the speedup of parallel processing. This aspect of work load balancing is discussed in more detail in chapter 3.

The Block Comparison Technique using Sample-by-Sample comparison method cannot be used for grouping of blocks with equal intensity values. Therefore, we can say that the Block Comparator Technique using Divide and Conquer method is more suitable for parallel processing.

Table 2.7 SIF table for the Selection Sort method

Number of Blocks in an image (NB)	% of Number of Similar Blocks (NSB)	Number of Blocks in Match List	Total number of Base Operations for Non-Block Comparator Technique (T_{NBCT})	Number of Base Operations for Block Comparator Technique (T_{BCT})	Speed Improvement Factor (SIF)
256	0	0	942080	991104	0.951
	10%	26	942080	897088	1.050
	30%	77	942080	712672	1.322
	50%	128	942080	528256	1.783
	75%	192	942080	296832	3.174
	85%	218	942080	202816	4.645
	90%	230	942080	159424	5.909
	95%	244	942080	108800	8.659
	100%	256	942080	69088	13.636
4266	0	0	15698880	25069149	0.626
	10%	427	15698880	23525117	0.667
	30%	1280	15698880	20440669	0.768
	50%	2133	15698880	17356221	0.905
	75%	3200	15698880	13497949	1.163
	85%	3626	15698880	11957533	1.313
	90%	3839	15698880	11187325	1.403
	95%	4053	15698880	10413501	1.508
	100%	4266	15698880	9646973	1.627
15594	0	0	57385920	1.8E+08	0.319
	10%	1560	57385920	1.74E+08	0.329
	30%	4679	57385920	1.63E+08	0.352
	50%	7797	57385920	1.52E+08	0.378
	75%	11696	57385920	1.38E+08	0.417
	85%	13255	57385920	1.32E+08	0.435
	90%	14034	57385920	1.29E+08	0.444
	95%	14815	57385920	1.26E+08	0.454
	100%	15594	57385920	1.24E+08	0.464

Table 2.8 SIF table for the Divide and Conquer Sort method

Number of Blocks in an image (NB)	% of Number of Similar Blocks (NSB)	Number of Blocks in Match List	Total number of Base Operations for Non-Block Comparator Technique (T_{NBCT})	Number of Base Operations for Block Comparator Technique (T_{BCT})	Speed Improvement Factor (SIF)
256	0	0	942080	960512	0.981
	10%	26	942080	866496	1.087
	30%	77	942080	682080	1.381
	50%	128	942080	497664	1.893
	75%	192	942080	266240	3.538
	85%	218	942080	172224	5.470
	90%	230	942080	128832	7.312
	95%	244	942080	78208	12.046
	100%	256	942080	38496	24.472
4266	0	0	15698880	16023346	0.980
	10%	427	15698880	14479311	1.084
	30%	1280	15698880	11394863	1.378
	50%	2133	15698880	8310415	1.889
	75%	3200	15698880	4452143	3.526
	85%	3626	15698880	2894416	5.424
	90%	3839	15698880	2124208	7.390
	95%	4053	15698880	1350384	11.625
	100%	4266	15698880	601170	26.114
15594	0	0	57385920	58601129	0.979
	10%	1560	57385920	52960180	1.084
	30%	4679	57385920	41681876	1.377
	50%	7797	57385920	30407188	1.887
	75%	11696	57385920	16308404	3.519
	85%	13255	57385920	10578608	5.425
	90%	14034	57385920	7761744	7.393
	95%	14815	57385920	4937648	11.622
	100%	15594	57385920	2216905	25.886

Table 2.9 SIF table for the Sample-by-Sample comparison method

Number of Blocks in an image (NB)	% of Number of Similar Blocks (NSB)	Number of Blocks in Match List	Total number of Base Operations for Non-Block Comparator Technique (T_{NBCT})	Number of Base Operations for Block Comparator Technique (T_{BCT})	Speed Improvement Factor (SIF)
256	0	0	942080	958464	0.983
	10%	26	942080	862784	1.092
	30%	77	942080	675104	1.395
	50%	128	942080	487424	1.933
	75%	192	942080	251904	3.740
	85%	218	942080	156224	6.030
	90%	230	942080	112064	8.407
	95%	244	942080	60544	15.560
	100%	256	942080	20064	46.954
4266	0	0	15698880	15971904	0.983
	10%	427	15698880	14400544	1.090
	30%	1280	15698880	11261504	1.394
	50%	2133	15698880	8122464	1.933
	75%	3200	15698880	4195904	3.741
	85%	3626	15698880	2628224	5.973
	90%	3839	15698880	1844384	8.512
	95%	4053	15698880	1056864	14.854
	100%	4266	15698880	276704	56.735
15594	0	0	57385920	5.84E+07	0.983
	10%	1560	57385920	5.26E+07	1.090
	30%	4679	57385920	4.12E+07	1.394
	50%	7797	57385920	2.97E+07	1.933
	75%	11696	57385920	1.53E+07	3.740
	85%	13255	57385920	9605536	5.974
	90%	14034	57385920	6738816	8.516
	95%	14815	57385920	3864736	14.849
	100%	15594	57385920	1.00E+06	57.289

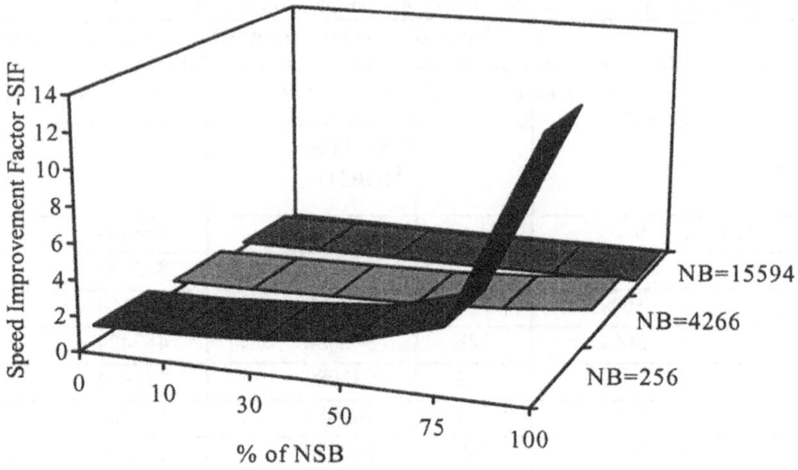

Figure 2.19 SIF Vs NSB for the Selection Sort method

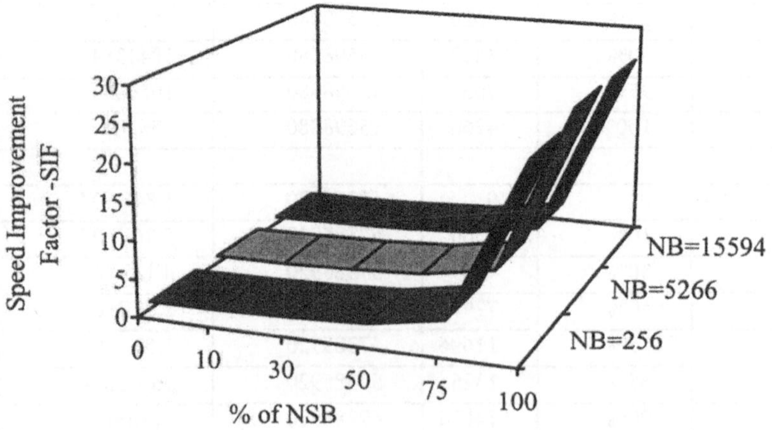

Figure 2.20 SIF Vs NSB for the Divide and Conquer Sort method

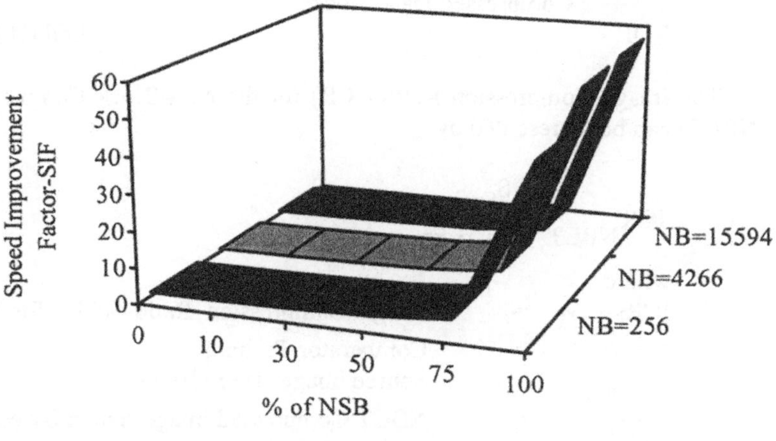

Figure 2.21 SIF Vs NSB for the Sample-by-Sample Comparison method

2.4.2 Comparison of the Non-Block Comparator Technique and Block Comparator Technique Image Compression Ratio

In this section the Image Compression Ratio obtained for the Digital Image Compression Techniques are calculated with and without the proposed Block Comparator Technique. Image Compression Ratio (ICR) is defined as the ratio of original image data size to the compressed image data size as given by,

$$ICR = \frac{S_{Srcimg}}{S_{Compimg}} \tag{2.20}$$

where,

ICR	=	Image Compression Ratio,
S_{Srcimg}	=	Source image data size in Bytes,
$S_{Compimg}$	=	Compressed image data size in Bytes.

Image Compression Ratio for the Non-Block Comparator Technique is given in section 2.4.2.1. Image Compression Ratio for the Block Comparator Technique is given in section 2.4.2.2. A comparison of the Image Compression Ratios obtained by these techniques is given in section 2.4.2.3.

2.4.2.1 Image Compression Ratio for the Non-Block Comparator Technique. In this section the formula is given for the Image Compression Ratio for the Non-Block Comparator Technique. With the standard Non-Block Comparator Technique, the compressed image data structure can be represented in the following format, as discussed in section 2.3.4.

JHI		JPEG Header Information
	Compressed Data Blocks	
EOI		End Of Image Marker

The Image Compression Ratio (ICR) for the Non-Block Comparator Technique (NBCT) can be represented by,

$$ICR_{NBCT} = \frac{S_{Srcimg}}{S_{NBCT}} \qquad (2.21)$$

where,

ICR_{NBCT} = Image Compression Ratio (ICR) for the Non-Block Comparator Technique,

S_{Srcimg} = Source image size in Bytes,

S_{NBCT} = NBCT Compressed image size in Bytes.

S_{NBCT} can be expanded further, as given in equation 2.22. Compressed block data size can be represented by the ratio of original image block size and the Block Compression Factor (BCF). Compression factor depends on the quality of compressed image, specified by the user.

$$S_{NBCT} = S_{JHI} + NB * \frac{S_{Blk}}{BCF} + S_{EOI} \qquad (2.22)$$

where,

S_{NBCT} = NBCT Compressed image size in Bytes,

S_{JHI} = Size of JPEG Header Information in Bytes,

NB = Number of Blocks in the image,

S_{Blk} = Size of one block in Bytes,

BCF = Block Compression Factor,

S_{EOI} = Size of End Of Image marker.

2.4.2.2 Image Compression Ratio for the Block Comparator Technique. In this section the formula is given for the Image Compression Ratio using the Block Comparator Technique (BCT). The Image Compression Ratio for Block Comparator Technique is given by,

$$ICR_{BCT} = \frac{S_{Srcimg}}{S_{BCT}} \qquad (2.23)$$

where,

ICR_{BCT} = Image Compression Ratio for Block Comparator Technique,

S_{Srcimg} = Source image size in Bytes,

S_{BCT} = BCT Compressed image size in Bytes.

For the Block Comparator Technique, the Compressed Image Data Structure can be represented in many formats. Three different structures were selected for the analysis. The Compressed image size for these Compressed Image Data Structures are given in the following sections.

In all of these storage formats the compressed image blocks are stored in two distinct groups. The first group consists of blocks that match with a unique block. This group is called as the Similar Block Group (SBG).

The second group consists of all the unique blocks. This group is called as the Unique Block Group (UBG). Then the data structure for the compressed image will be as follows:

JHI	JPEG Header Information
SBG	Similar Block Group
UBG	Unique Block Group
EOI	End Of Image marker

Compressed image size for Compressed Image Data structure-1: For the Block Comparator Technique, the Compressed Image Data Structure-1 (CIDS-1) is given in table 2.10. The JPEG Header Information (JHI) is the first item for the Non-Block Comparator Technique. Following the header the Similar Block Group is stored. The first component of this block is stored following the Similar Block Marker. In this group each block is stored as an ordered pair comprising; Block Number followed by the matching Unique Block Number. After the Similar Block Group comes the Unique Block Group, where the first component is the Unique Block Marker. In the Unique Block Group all unique blocks are stored in the correct sequence. The original block number for each Unique Block can be identified by looking up the missing block numbers in the Similar Block Group. The first block is always taken as Unique Block-1. If it turns out that Block-2 and Block-3 match with Block-1 then these will be stored in the Similar Block Group. If Block-4 is the next unique block then this block will be the second block stored in the Unique Block Group and Block-4 will be missing from the Similar Block Group. The End Of Image (EOI) marker is stored at the end of the file.

The size of the JPEG information header and the EOI marker are same as in the JPEG compressed image. The size of block number is taken as sixteen bits because in large images the number of blocks are more than 256. Size for various markers is taken as eight bits. The size of the Compressed Image Data Structure-1 for the Block Comparator Technique can be written as,

$$S_{BCT1} = S_{JHI} + S_{SBM} + NSB * (S_{BNF} + S_{UBNF}) + S_{UBM} + NUB * \frac{S_{Blk}}{BCF} + S_{EOI} \tag{2.24}$$

where,

S_{BCT1}	=	BCT Compressed image size for CIDS-1 in Bytes,
S_{JHI}	=	Size of the JPEG Header Information in Bytes,
S_{SBM}	=	Size of the Similar Block Marker,
NSB	=	Number of Similar Blocks,
S_{BNF}	=	Size of the Block Number Field,
S_{UBNF}	=	Size of the Unique Block Number Field,
S_{UBM}	=	Size of the Unique Block Marker,
NUB	=	Number of Unique Blocks,
S_{Blk}	=	Size of one block in Bytes,
BCF	=	Block Compression Factor,
S_{EOI}	=	Size of the End Of Image marker.

Table 2.10 Compressed Image Data Structure-1

Component	Parameter	Symbol	Size (Bytes)
JHI	JPEG Information Header	JHI	
SBG	Similar Block Marker	SBM	1
	Block Number, Unique Block No	BN, UBN	2 + 2
	Block Number, Unique Block No	BN, UBN	2 + 2
	....		
UBG	Unique Block Marker	UBM	1
	Unique Block Compressed data	UBCD	S_{Blk} /CF
	....		
EOI	End of Image marker	EOI	1

Compressed image size for Compressed Image Data Structure-2: For the Block Comparator Technique, the Compressed Image Data Structure-2 (CIDS-2) is given in table 2.11. The CIDS-2 is similar to the CIDS-1 except for the storage format of the compressed data in the Unique Block Group. In CIDS-2 the compressed data for each Unique Block is also stored as an ordered pair: Unique Block Number followed by the Unique Block Compressed Data.

The size of each field is the same as in the Compressed Image Data Structure-1. Size of the Unique Block Number Field is taken as sixteen bits. The size of CIDS-2 for the Block Comparator Technique can be written as,

Size of the Unique Block Number Field is taken as sixteen bits. The size of CIDS-2 for the Block Comparator Technique can be written as,

$$S_{BCT2} = S_{JHI} + S_{SBM} + NSB * (S_{BNF} + S_{UBNF}) + S_{UBM} + NUB * (S_{UBNF} + \frac{S_{Blk}}{BCF}) + S_{EOI} \qquad (2.25)$$

where,

S_{BCT2} = BCT Compressed image size for CIDS-2 in Bytes,
S_{JHI} = Size of JPEG Header Information in Bytes,
S_{SBM} = Size of Similar Block Marker,
NSB = Number of Similar Blocks,
S_{BN} = Size of the Block Number Field,
S_{UBN} = Size of the Unique Block Number Field,
S_{UBM} = Size of the Unique Block Marker,
NUB = Number of Unique Blocks,
S_{Blk} = Size of one block in Bytes,
BCF = Block Compression Factor,
S_{EOI} = Size of the End Of Image marker.

Table 2.11 Compressed Image Data Structure-2

Component	Parameter	Symbol	Size (Bytes)
JHI	JPEG Information Header	JHI	
SBG	Similar Block Marker	SBM	1
	Block Number, Unique Block No	BN, UBN	2 + 2
	Block Number, Unique Block No	BN, UBN	2 + 2
	....		
UBG	Unique Block Marker	UBM	1
	Unique Block Number, Block Compressed data	UBN, BCD	2, S_{Blk} /CF
	Unique Block Number, Block Compressed data	UBN, BCD	2, S_{Blk} /CF
	....		
EOI	End of Image marker	EOI	1

Compressed image size for Compressed Image Data Structure-3: The Compressed Image Data Structure-3 (CIDS-3) is given in table 2.12. In CIDS-3 the JPEG Header Information, Unique Block Group (UBG) and EOI marker are the same as for CIDS-1. Following the header the Similar Block Group (SBG) is stored. The SBG consists a number of Similar Block Lists (SBL). Each SBL starts with a Similar Block Marker and a Unique Block Number followed by the list of similar Block Numbers. The Similar Block Marker at the head of the next Similar Block list also marks the end of the previous Similar Block list.

The size of CIDS-3 for Block Comparator Technique can be written as,

$$S_{BCT3} = S_{JHI} + NL * (S_{SBM} + S_{UBNF} + nl * S_{BNF}) + S_{UBM} + NUB * \frac{S_{Blk}}{BCF} + S_{EOI} \quad (2.26)$$

where,

S_{BCT3} = BCT Compressed image size for CIDS-3 in Bytes,
S_{JHI} = Size of the JPEG Header Information in Bytes,
NL = Number of Similar Block Lists in SBG,
nl = Number of Block Numbers, followed in each Unique Block list,
S_{SBM} = Size of the Similar Block Marker,
S_{UBNF} = Size of the Unique Block Number Field,
S_{UBM} = Size of the Unique Block Marker,
NUB = Number of Unique Blocks,
S_{Blk} = Size of one block in Bytes,
S_{EOI} = Size of the End Of Image marker.

Table 2.12 Compressed Image Data Structure-3

Component	**Parameter**	**Symbol**	**Size (Bytes)**
JHI	JPEG Information Header	JHI	
SBG	Similar Block Marker	SBM	1
	Unique Block Number, Block No, Block No, Block No, ...	UBN, BN, BN ..	2 + 2 + 2 ...
	Similar Block Marker	SBM	1
	Unique Block Number, Block No, Block No, Block No, ...	UBN, BN, BN ...	2 + 2 + 2 ...
UBG	Unique Block Marker	UBM	1
	Unique Block Compressed data	UBCD	S_{Blk} /CF
EOI	End of Image marker	EOI	1

2.4.2.3 Comparison of Image Compression Ratios. Image Compression Ratio (ICR) for Non-Block Comparator Technique is given in table 2.13 and graph of the same is shown in figure 2.22. Image Compression Ratio for Block Comparator Technique for three Compressed Image Data Structures (CIDS-1, CIDS-2 and CIDS-3) are given in tables 2.14, 2.15, 2.16 and 2.17. In each of these tables, the values of Image Compression Ratio are calculated for three image sizes, viz. 125 x 125, 625 x 423, and 1100 x 900. For each image size, the values of ICR are calculated for different output image qualities, viz. 100, 75, 50 and 25 percent. For CID-3, the graph of ICR Vs % quality for NSB = 10%, 30%, 50%, and 75% are shown in figure 2.23, 2.24, 2.25 and 2.26 respectively.

Conclusions are derived by comparing ICR values from all Compressed Image Data Structures (CIDS) and by comparing ICR values from each CIDS separately. The conclusions derived are given in the following sections.

Conclusions from all CIDS: From tables 2.13 to 2.17 we can see that the ICRs increase as the image size increases irrespective of the quality of the output image. For each image size the ICR increases with decrease in quality of the output image.

Conclusion for CIDS-1: From table 2.13 and tables 2.14 to 2.17 we can see that for the CIDS-1 and quality = 100 Image Compression Ratio is slightly greater than the same for the Non-Block Comparator Technique. For quality = 75 Image Compression Ratio is almost equal to the same for Non-Block Comparator Technique. For quality = 50 and 25 Image Compression Ratio values are less compared to the same for the Non-Block Comparator Technique. This indicates that the size of the compressed image using Compressed Image Data Structure-1 for Block Comparator Technique is beneficial for image quality greater than 75% only.

Conclusions for CIDS-2: By comparing the ICR values of CIDS-2 and CIDS-1, we can see that ICR values for CIDS-2 are less than those for CIDS-1 for all image sizes. This is because the Unique Block Number is stored in the Unique Block Group. This data structure is more robust than the CIDS-1 data structure, because all the blocks numbers are included in the data structure.

By comparing Image Compression Ratio values of CIDS-2 and Non-Block Comparator Technique, we can see that the Image Compression Ratio values of CIDS-2 for quality = 100 are almost equal to that of Non-Block Comparator Technique. For quality less than 100, Image Compression Ratio values are less than the same for Non-Block Comparator Technique.

Conclusions for CIDS-3: By comparing Image Compression Ratio values of CIDS-3 with those for the other two data structures, we can see that the ICR values for CIDS-3 are greater in all cases.

By comparing CIDS-3 with Non-Block Comparator Technique, we can see that the Image Compression Ratio values for CIDS-3 for quality = 100 and 75 are greater than the same for Non-Block Comparator Technique. For quality = 50 the values are almost equal. This indicates that the CIDS-3 data structure is better than the others for quality greater than 50.

By comparing all three Compressed Image Data Structures, we can say that CIDS-3 is the best of these three. Therefore, CIDS-3 data structure is chosen to measure the improvement over the Non-Block Comparator Technique.

Image Compression Ratio Improvement Factor (ICRIF): The improvement in the compression ratio is represented by a factor called the Image Compression Ratio Improvement Factor (ICRIF). ICRIF can be defined as the ratio of Image Compression Ratio for Block Comparator Technique to the Image Compression Ratio for Non-Block Comparator Technique, as given by,

$$ICRIF = \frac{ICR_{BCT}}{ICR_{NBCT}} \tag{2.27}$$

where,
ICR_{BCT} = Image Compression Ratio for the Block Comparator Technique,
ICR_{NBCT} = Image Compression Ratio for the Non-Block Comparator Technique.

Table 2.18 shows the ICRIF for CIDS-3 using NSB = 75 graph of ICRIF versus quality is shown in figure 2.27. From table we can see that the Image Compression Ratio Improvement Factor (ICRIF) is almost equal for all image sizes irrespective of the quality of output image. There is no benefit in using the Block Comparator Technique for images with less than 50% quality. By using the Block Comparator Technique we can get an improvement of 2.8 times over the Non-Block Comparator Technique for quality = 100.

Table 2.13 Image Compression Ratio for Non-Block Comparator Technique

Number of Blocks NB	Quality %	Compressed image size for NBCT: S_{NBCT}	Image Compression Ratio for NBCT: ICR_{NBCT}
256	100	3450	4.529
	75	993	15.735
	50	642	24.338
	25	501	31.188
4266	100	54778	4.895
	75	13825	19.394
	50	7974	33.625
	25	5634	47.591
15594	100	199777	4.956
	75	50074	19.771
	50	28688	34.509
	25	20134	49.171

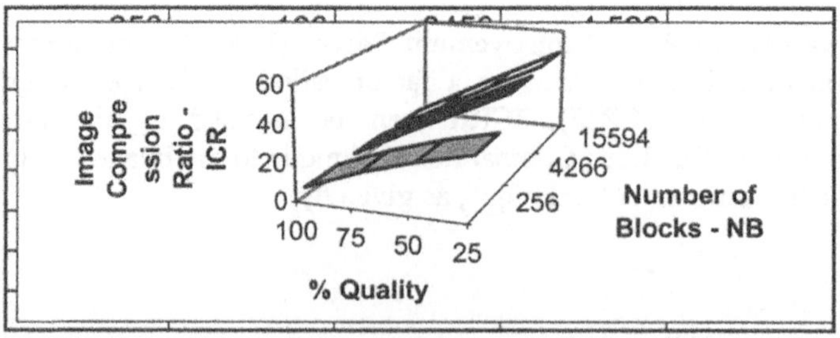

Figure 2.22 ICR Vs quality for NBCT

Table 2.14 Image Compression Ratio for Block Comparator Technique using three Compressed Image Data Structures for Number of Similar Block = 10%

Number of Blocks NB	Quality %	Image Compression Ratio for CIDS-1 (ICR_{BCT1})	Image Compression Ratio for CIDS-2 (ICR_{BCT2})	Compressed image size for CIDS-3 (S_{BCT3})	Image Compression Ratio for CIDS-3 (ICR_{BCT3})
256	100	4.846	4.241	3176	4.920
	75	15.379	10.586	968	16.142
	50	22.321	13.470	648	24.113
	25	27.221	15.111	522	29.933
4266	100	5.255	4.568	50179	5.343
	75	18.925	12.273	13324	20.123
	50	30.116	16.171	8051	33.303
	25	39.448	18.523	5945	45.101
15594	100	5.321	4.624	182939	5.412
	75	19.289	12.470	48212	20.534
	50	30.862	16.460	28966	34.178
	25	40.609	18.876	21267	46.551

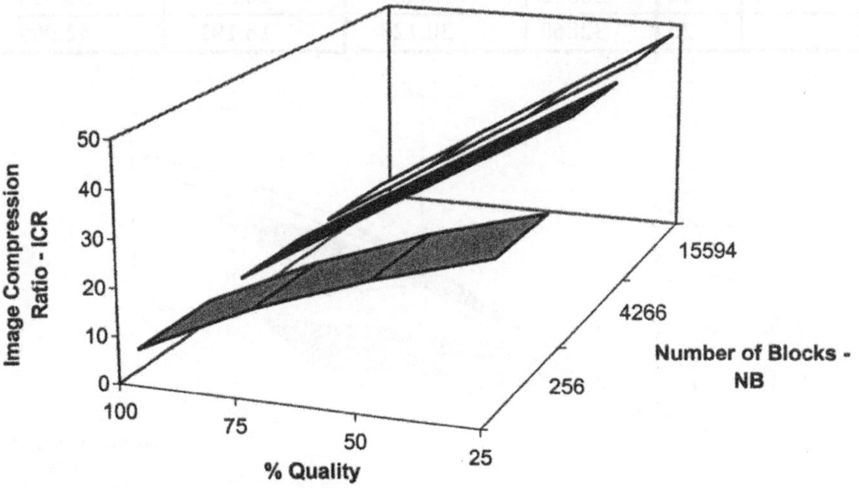

Figure 2.23 ICR Vs quality for CIDS-3 (NSB = 10%)

Table 2.15 Image Compression Ratio for Block Comparator Technique using three Compressed Image Data Structures for Number of Similar Block = 30%

Number of Blocks NB	Quality %	Compressed image size for CIDS-1 (S_{BCT1})	Image Compression Ratio for CIDS-1 (ICR_{BCT1})	Image Compression Ratio for CIDS-2 (ICR_{BCT2})	Image Compression Ratio for CIDS-3 (ICR_{BCT3})
256	100	2775	5.631	4.987	5.930
	75	1056	14.796	11.050	17.058
	50	811	19.266	13.366	23.286
	25	713	21.914	14.589	27.269
4266	100	43516	6.162	5.418	6.545
	75	14851	18.054	12.876	21.801
	50	10756	24.928	16.029	32.714
	25	9118	29.406	17.768	40.885
15594	100	158612	6.242	5.486	6.632
	75	53819	18.395	13.086	22.260
	50	38848	25.484	16.315	33.555
	25	32860	30.128	18.101	42.099

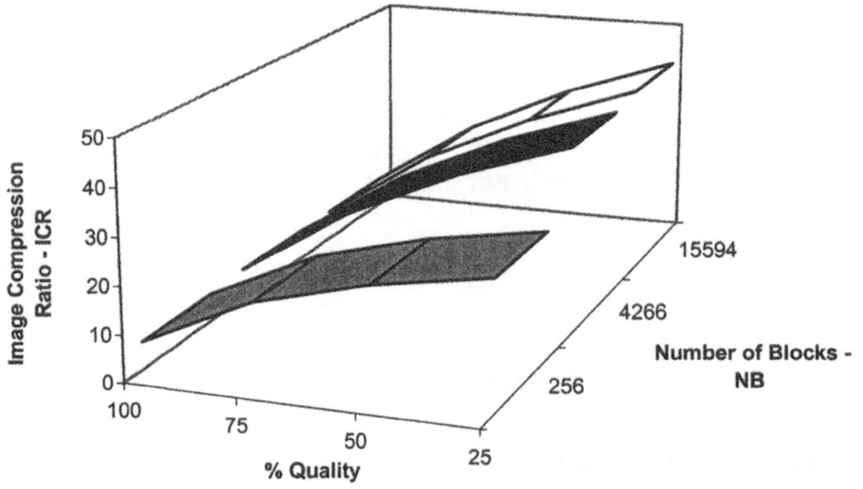

Figure 2.24 ICR Vs quality for CIDS-3 (NSB = 30%)

Table 2.16 Image Compression Ratio for Block Comparator Technique using three Compressed Image Data Structures for Number of Similar Block = 50%

Number of Blocks NB	Quality %	Image Compression Ratio for CIDS-1 (ICR_{BCT1})	Image Compression Ratio for CIDS-2 (ICR_{BCT2})	Compressed image size for CIDS-3 (S_{BCT3})	Image Compression Ratio for CIDS-3 (ICR_{BCT3})
256	100	6.718	6.052	2078	7.519
	75	14.243	11.548	849	18.404
	50	16.947	13.264	674	23.182
	25	18.361	14.115	603	25.912
4266	100	7.446	6.657	31758	8.443
	75	17.262	13.542	11281	23.768
	50	21.266	15.890	8348	32.118
	25	23.442	17.074	7178	37.354
15594	100	7.548	6.746	115585	8.565
	75	17.580	13.768	40734	24.304
	50	21.701	16.173	30041	32.955
	25	23.945	17.387	25764	38.426

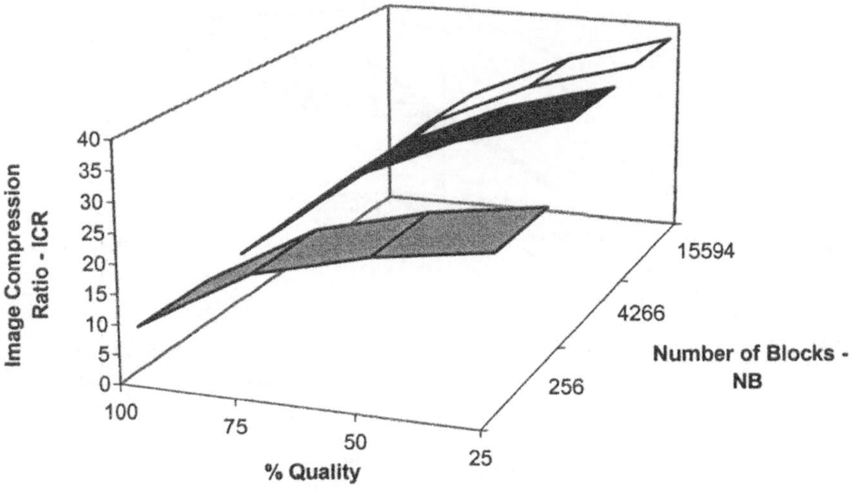

Figure 2.25 ICR Vs quality for CIDS-3 (NSB = 50%)

Table 2.17 Image Compression Ratio for Block Comparator Technique using three Compressed Image Data Structures for Number of Similar Block = 75%

Numbe r of Blocks NB	Qual ity %	Image Compression Ratio for CIDS-1 (ICR_{BCT1})	Image Compression Ratio for CIDS-2 (ICR_{BCT2})	Compres sed image size for CIDS-3 (S_{BCT3})	Image Compression Ratio for CIDS-3 (ICR_{BCT3})
256	100	8.863	8.263	1387	11.265
	75	13.611	12.245	772	20.240
	50	14.727	13.141	685	22.810
	25	15.244	13.552	649	24.076
4266	100	10.072	9.325	20228	13.255
	75	16.362	14.478	9995	26.826
	50	17.965	15.719	8525	31.452
	25	18.698	16.278	7940	33.769
15594	100	10.222	9.460	73470	13.475
	75	16.657	14.726	36049	27.463
	50	18.304	15.998	30703	32.244
	25	19.057	16.570	28565	34.658

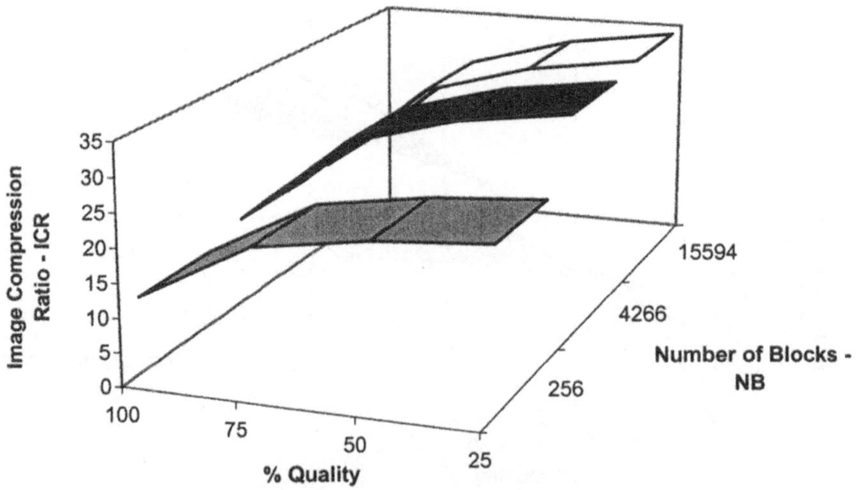

Figure 2.26 ICR Vs quality for CIDS-3 (NSB = 75%)

Table 2.18 ICRIF for the BCT using the Compressed Image Data Structures-3 for NSB = 75%

Number of Blocks NB	Quality %	Image Compression Ratio for NBCT (ICR_{NBCT})	Image Compression Ratio for CIDS-3 (ICR_{BCT3})	Image Compression Ratio Improvement Factor ICRIF
256	100	4.529	11.265	2.487
	75	15.735	20.240	1.286
	50	24.338	22.810	0.937
	25	31.188	24.076	0.772
4266	100	4.895	13.255	2.708
	75	19.394	26.826	1.383
	50	33.625	31.452	0.935
	25	47.591	33.769	0.710
15594	100	4.956	13.475	2.719
	75	19.771	27.463	1.389
	50	34.509	32.244	0.934
	25	49.171	34.658	0.705

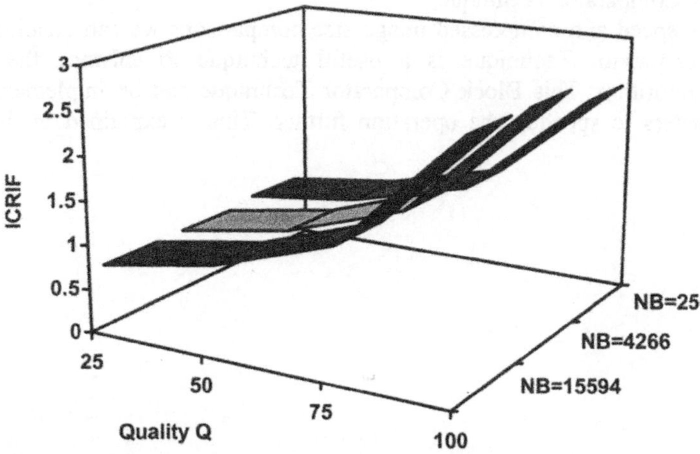

Figure 2.27 ICRIF Vs quality

2.5 Summary

Various digital image compression techniques such as Vector Quantisation (VQ), fractal, wavelet and JPEG were explained in this chapter. The JPEG technique is used in many applications. Therefore the JPEG algorithm is chosen for our research purpose. The JPEG algorithm can be implemented in hardware as well as software. Many of the limitations of hardware implementation can be overcome in software based implementations.

In the JPEG algorithm all image blocks are processed individually. These blocks of compressed image are stored sequentially. In many types of images, there is the possibility of having one or more similar blocks in the image. Improvement in performance of the compression algorithm can be achieved by locating similar blocks in the image. The Block Comparator Technique was proposed to enhance the performance of the JPEG algorithm. An overview of the Block Comparator Technique was given in this chapter.

The Block Comparator Technique improves the speed of the compression and decompression operations and reduces the size of the compressed data file. These two factors have been discussed in this chapter. Improvement in the speed of Block Comparator Technique is expressed in terms of Speed Improvement Factor. Using analytical methods, the Speed Improvement Factor for the Block Comparator Technique using Divide and Conquer method was found to be suitable for all image sizes.

Three types of Image Data Structures were introduced for the Block Comparator Technique. By comparing the three Image Data Structures, it was concluded that the Image Data Structure-3 is more suitable for all image sizes. The Image Compression Ratio improvement Factor of Image Data Structure-3 $\cong$ 2.8. This indicates that the compressed image size for CIDS-3 is 2.8 times less than the compressed image size of the Non-Block Comparator Technique.

From these speed and compressed image size comparisons we can conclude that the Block Comparator Technique is a useful technique to enhance the JPEG compression algorithm. This Block Comparator Technique can be implemented on parallel computers to speedup the operation further. This is explained in the next chapter.

3 PARALLEL PROCESSING PLANS FOR DIGITAL IMAGE COMPRESSION TECHNIQUES

3.1 Introduction

This chapter describes methods used for parallel processing of digital image compression algorithms. Various types of parallel computers and parallel processing Plans for digital image compression are described.

Parallel computers can be classified based on memory access technique, network topology and some other issues. Three parallel computers were chosen for the experimentation, each with a different memory access architectures, viz. the Mercury system with a distributed memory architecture, the Shiva system with a shared memory and the Param system with a hybrid memory architecture.

Digital image compression can be implemented in a variety of ways on parallel computers. Each uniquely identifiable way of implementation is called as a 'Plan'. Some of these Plans were implemented on available parallel computers and other Plans were simulated using the Network II.5 simulation package. Performance of these Plans can be evaluated in terms of speedup, scaleup, and efficiency.

Parallel processing is often used to increase the computation speed of a task by dividing the task into several sub-tasks and allocating multiple processors to execute multiple sub-tasks simultaneously. A wide variety of parallel computers are used for parallel processing. Section 3.2 briefly describes the various parallel computer architectures and parallel programming languages.

Digital image compression can be implemented on parallel computers in a variety of ways. Each way of performing image compression is called as a 'Plan' in this book. Each Plan used for digital image compression on a parallel computer can be specified by a 6-tuple consisting of the image compression technique, block dependency, image partitioning method, memory architecture, memory organisation / network topology and the number of processors. Section 3.3 describes these Plans in detail. Some of these Plans were implemented on available parallel computers such as Mercury, Shiva and Param systems. Section 3.4.1 describes the implementation detail on these parallel computers. Other Plans were simulated using the Network II.5 simulation package. Simulation models for these Plans are described in section 3.4.2.

These plans were evaluated in terms of speedup, scaleup, and efficiency. These terms are defined in section 3.5.

3.2 Parallel Computer Architectures

One of the main aims of this research project was to investigate the application of parallel computers architectures to speedup digital image compression operations. Parallel computer architectures can be classified based on factors described below [Krishnamurthy, 89].

1 **Granularity**: The number of processors used in a system is a measure of its granularity. Granularity can be classified as fine grain, medium grain, and coarse grain. In a fine grained parallel computer several thousand processors are used. In a medium grained parallel computer a few tens to several hundred processors may be used. Whereas, in a coarse grained parallel computer only a few processors are used.

2 **Interconnection topology:** The processors can be interconnected to form a network with topologies such as Ring, Array, Mesh, Tree, Cube, Pyramid etc.

3 **Task allocation**: Task allocation on any architecture is a mapping of the program onto the available machine resources. This can be done statically or dynamically. In a static mapping all the tasks are allocated prior to program execution, based on the system topology. In dynamic allocation, the tasks may be migrated from one processor to another to balance the work load.

One of the most widely used taxonomy for parallel architecture was proposed by Flynn. In this taxonomy singularity or multiplicity of instruction stream and data stream is the basis for classification. This gives us four possible classes of parallel computers, namely [Krishnamurthy, 89],

1 Single Instruction stream Single Data stream (SISD) machines,
2 Single Instruction stream Multiple Data stream (SIMD) machines,
3 Multiple Instruction stream Single Data stream (MISD) machines,
4 Multiple Instruction stream Multiple Data stream (MIMD) machines.

A SIMD machine is a computer system consisting of a single control unit, N processors, M memory modules, and an interconnections network. The instructions are broadcast from the control unit to all the processors and processors execute the same instructions at the same time [Siegel, 85].

MIMD architectures operate in parallel in an asynchronous manner and generally share data either through a common memory or by passing data over communication channels. Some of the commercially available MIMD computers are CM-5, NCUBE, iWarp, iPSC, Paragon, Meiko computing system, Teradata etc. [Hord, 93].

Based on memory architecture parallel computers are classified as shared memory, distributed memory, and hybrid memory architectures. These are explained in the following sections.

3.2.1 Shared Memory Architecture

In a shared memory architecture the processors use a shared memory space for passing data and messages. Shared memory architecture is further divided into global memory architecture and global-plus-local memory organisation. In the global memory organisation, there is a only one main memory module and all processors access the same global memory. Whereas, in a global-plus-local memory organisation each processor has its own local memory, and can also access the main global memory. These two types of architectures are shown in figure 3.1.

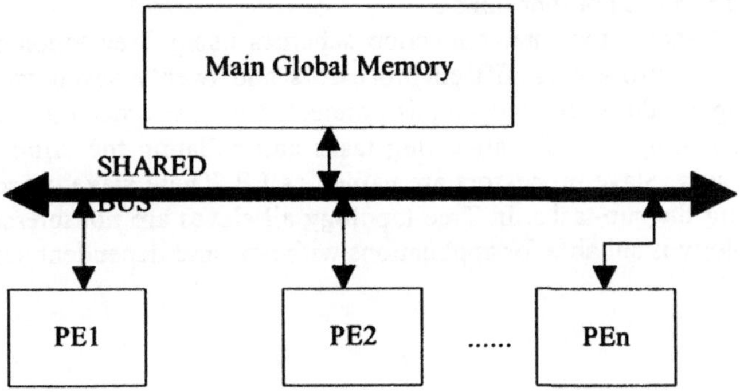

(a)

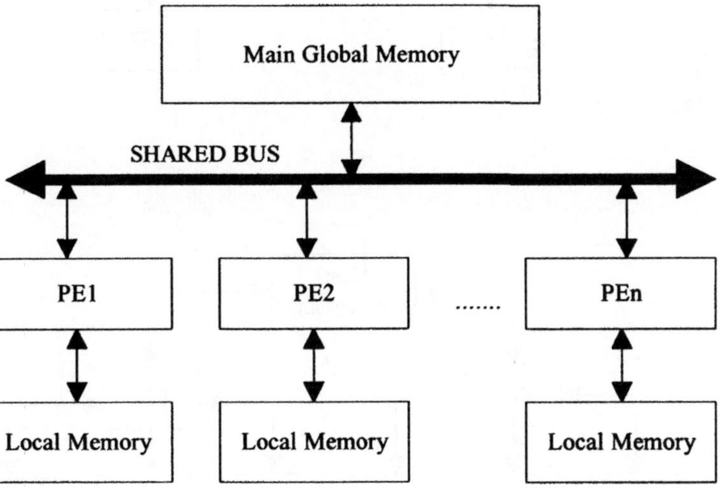

(b)

Figure 3.1 Shared Memory Architectures

a) with global memory organisation b) with global-plus-local memory organisation

3.2.2 Distributed Memory Architecture

In the distributed memory architecture all processors have only their own local memory. Processors can be connected in many different interconnection topologies. Based upon the interconnection topology, distributed memory architectures can be classified as Tree, Mesh, Cube, and Pyramid architectures. These topologies with various number of processors are shown in figure 3.2, 3.3, 3.4 and 3.5. These figures show the interconnection schemes with each processor having upto four links. It is easy to construct these configurations in a Transputer based system, since each Transputer T805 has got four links.

Figure 3.2 shows the interconnection schemes using Tree topologies with three processors, nine processors, fifteen processors and twenty seven processors. In all these topologies, the host processor is connected to two other processors. The host processor is mainly used for allocating tasks and collating the information from all other processors. Slave processors are named as PE. These slave processors are used for processing the sub-tasks. In Tree topology all slaves are not interconnected. This type of topology is suitable for applications with non-interdependent tasks.

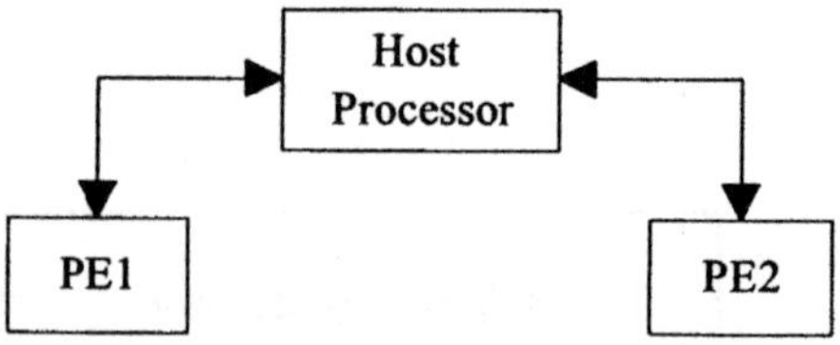

(a)

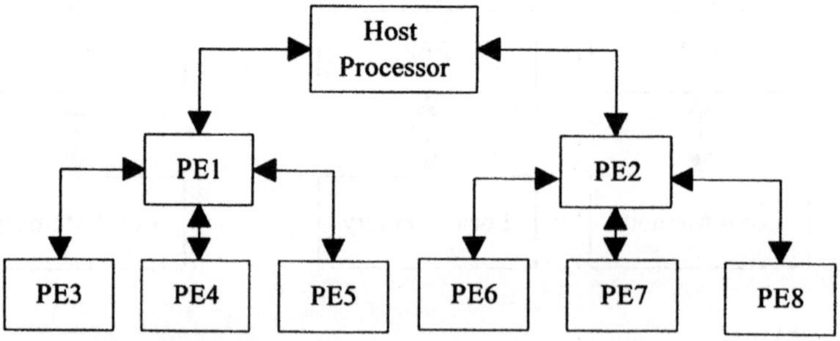

(b)

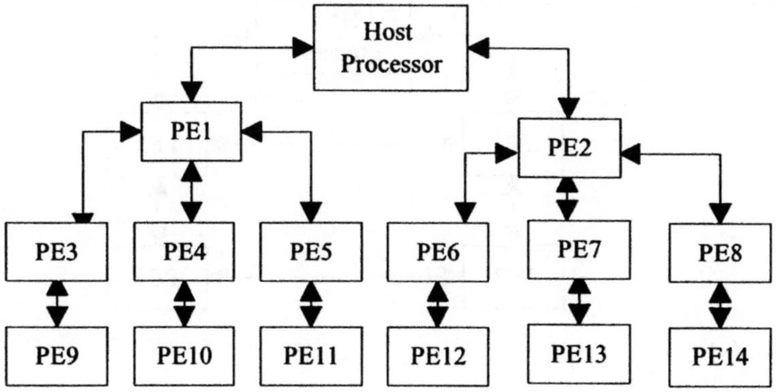

(c)

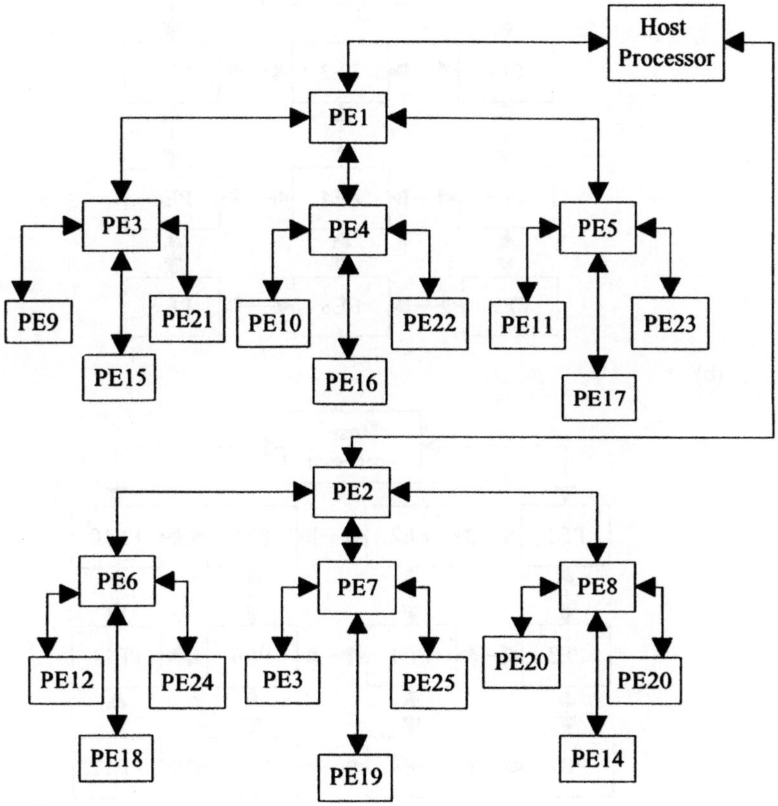

(d)

Figure 3.2 Tree topologies with

(a) three processors (b) nine processors
(c) fifteen processors (d) twenty seven processors

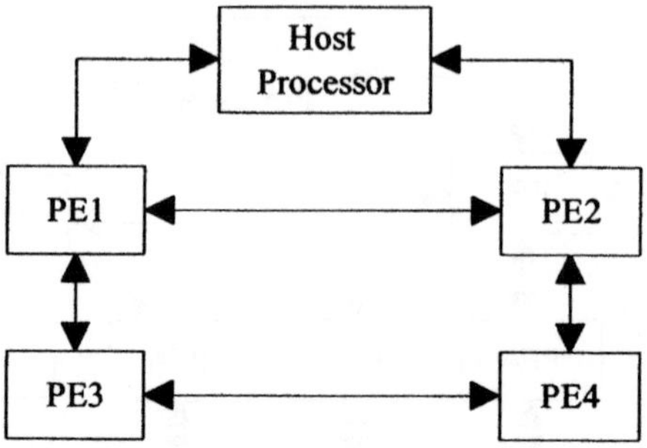

(a)

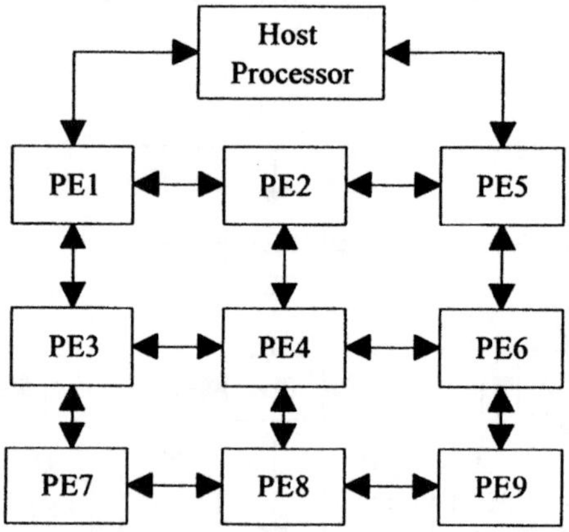

(b)

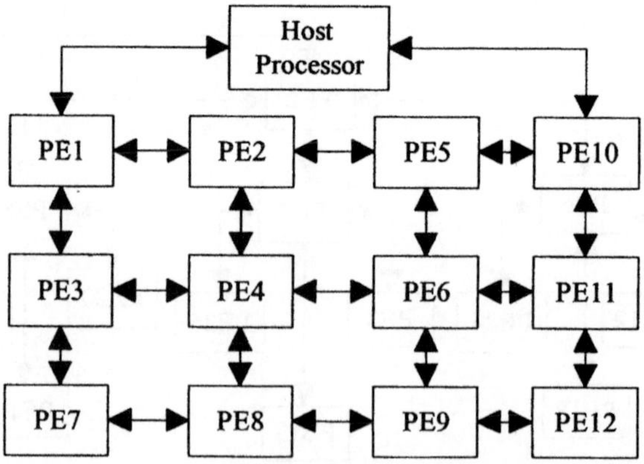

(c)

Figure 3.3 Mesh topologies with
(a) five processors (b) ten processors (c) seventeen processors

Figure 3.3 shows the interconnection schemes using Mesh topologies with five processors, ten processors, and seventeen processors. The host processor in these topologies is connected to two slave processors. All slave processors are interconnected in A Mesh structure. This type of Mesh topology is mainly used for applications which have interdependent tasks. The Mesh topology can easily be converted into a Torus topology. In Torus topology last column processors are connected to the first column processors and last row processors are connected to first row processors. Torus topology is often used in image processing applications.

Figure 3.4 shows the interconnection schemes using Pyramid topology with five processors and twenty one processors. In this topology, the host processor is connected to four slave processors. The twenty one processors topology is similar to a Quad-Tree topology where each processor is connected to four other processors. This topology is suitable for applications with both interdependent and non-interdependent task.

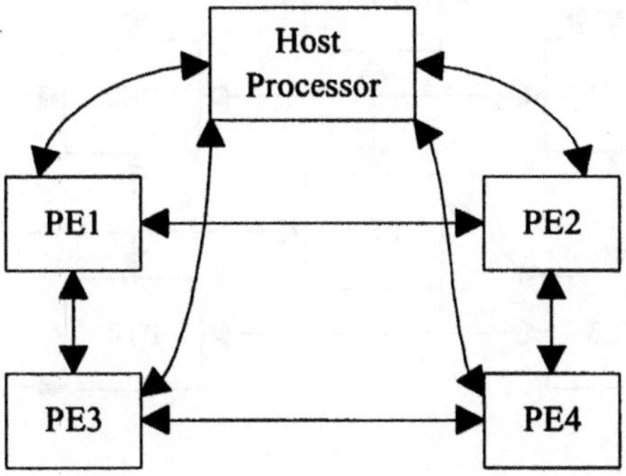

(a)

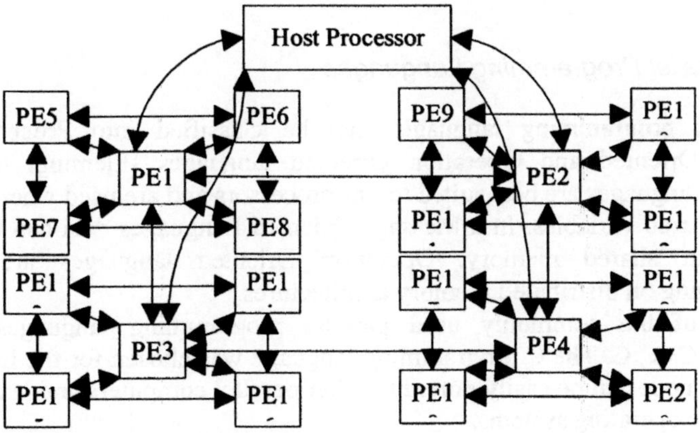

(b)

Figure 3.4 Pyramid topologies with

(a) five processors (b) twenty one processors

Figure 3.5 shows the 3-dimensional Cube architecture with nine processors, where host processor is connected to four slave processors.

Interconnection schemes described above are only indicative of the manner in which various topologies can be constructed. Different sets of interconnection schemes can be generated by varying the number of connection available of the processing elements.

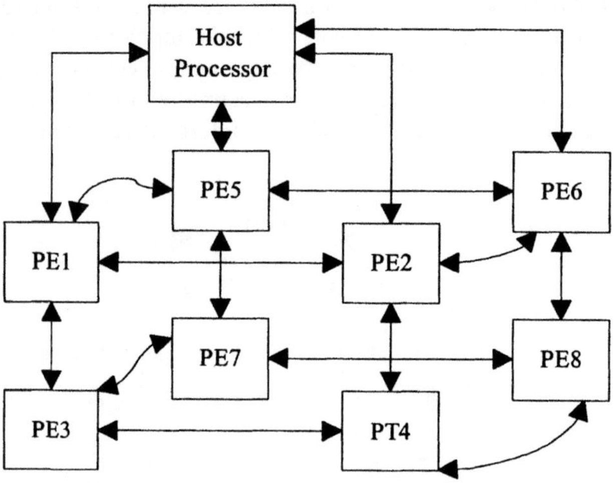

Figure 3.5 Cube topology

3.2.3 Parallel Programming Languages

Parallel programming languages can be classified into Procedure Oriented, Message Oriented and Operation Oriented languages [Fleming, 88]. Procedure Oriented Languages are best suited to uni-processors and are used when data is passed through shared variables. In a Message Oriented Languages data can be passed with or without shared memory. Operation Oriented languages are suitable for programming on distributed memory architectures.

Some of the commonly used parallel programming languages are Fortran, OCCAM, C++, C. The C programming language was chosen for the implementation, because C code can be easily ported to other parallel computers and it is supported by most of the operating systems.

3.3 Parallel Processing Plans for Digital Image Compression Techniques

Digital image compression technique can be implemented on parallel computers in many different ways. A specific way of implementation is called as a 'Plan'. Each Plan used for digital image compression on a parallel computer can be specified by a 6-tuple consisting of Image Compression Technique (ICT), Block Dependency (BD), Image Partitioning Method (IPM), Memory Architecture (MA), Memory Organisation / Network Topology (NT) and the Number of Processors (NP). Plan (P) thus can be represented as:

P(ICT, BD, IPM, MA, MO/NT, NP)

where,

P is Plan for implementation,
ICT is the Image Compression Technique used for image processing,
BD is Block Dependency used for image processing,
IPM is Image Partitioning Method used for image processing,
MA is Memory Architecture used for parallel processor,
MO/NT is Memory Organisation / Network Topology used for parallel processor,
NP is Number of Processors.

These factors are explained in the following sections.

3.3.1 Image Compression Technique (ICT)

Image Compression Technique can be classified into conventional image compression technique and block comparator enhancement to the JPEG technique. The conventional image compression technique is based on the JPEG algorithm without any block comparisons. The conventional technique is called as the Non-Block Comparator Technique (NBCT) in this book. The block comparator enhancement to the JPEG algorithm, based on block comparison and the JPEG algorithm, was explained in section 2.4. This technique is called as the Block Comparator Technique (BCT) in this book. The structure of classification of image compression technique is shown in figure 3.6.

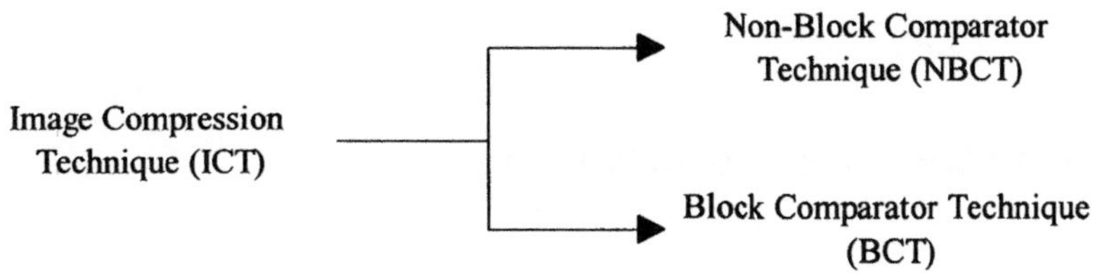

Figure 3.6 Classification of Image Compression Technique

3.3.2 Block Dependency (BD)

Most of the digital image compression algorithms use a block of 8 x 8 samples. To obtain a smoothing effect, the value of the neighbouring samples for each sample need to be considered. When the value of neighbouring samples are utilised the processor processing an image block, needs access to the edge samples in its neighbouring blocks. This is called Inter-Block Dependency (IBD) method. If smoothing effect is not required then samples in neighbouring blocks need not be accessed. This method is called as Non-Inter-Block Dependency (NIBD) method in this book.

On parallel computers, task allocation can be done in two ways in the Inter-Block Dependency method. In this method the basic blocks are called the Root Objects and edge sample of neighbouring blocks are called Leaf Objects. One way of allocating compression tasks is to allocate the task with only the Root Objects transmitted to the respective processors. Then the Leaf Objects (neighbouring samples) can be accessed from neighbouring processors. Since it requires run-time communication between processors it is called Inter-Processor Communication (IPC) method. The second way of task allocation is to allocate task with the Root Object along with the Leaf Objects. In this method there is no need of accessing neighbouring samples from other processors at run-time, eliminating the need for inter-processor communication. This method is called Non-Inter-Processors Communication (NIPC) method.

Classification of Block Dependency on parallel computers is shown in figure 3.7.

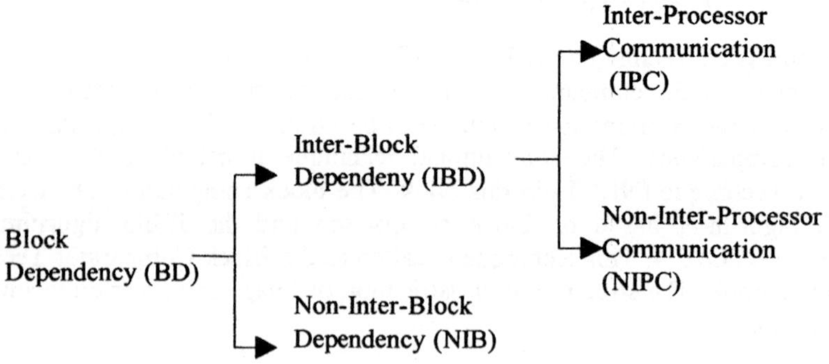

Figure 3.7 Classification of Block Dependency

3.3.3 Image Partitioning Method (IPM)

Most of the digital image compression algorithms use a block of 8 x 8 samples. Choice of an Image Partitioning Method based on blocks of an image is an important step, because it determines system efficiency, processing workload balance and the amount of usable parallelism.

Image Partitioning can be done in two ways. One method is to divide the complete image into a number of blocks. These blocks can be grouped into tasks equal to the number of processors. This method is called Block Based Image Partitioning (BBIP).

Second method of Image Partitioning can be done by dividing the complete image into a number of blocks such that each block is weighted in terms of intensity. The blocks can be grouped into a tasks which consists of equal intensity values. This method is called Balanced Workload Image Partitioning (BWIP).

Classification of Image Partitioning Method is shown in figure 3.8.

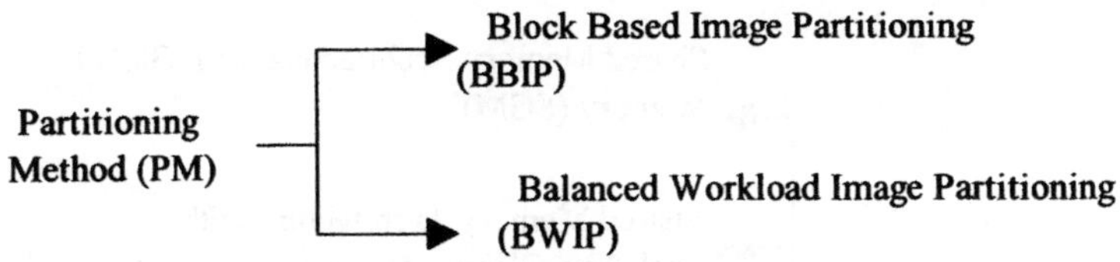

Figure 3.8 Classification of Image Partitioning Method

3.3.4 Memory Architecture (MA)

Based on memory access parallel computers are broadly classified as shared memory architecture, distributed memory architecture, and hybrid memory architecture as discussed in section 3.2. In this research shared and distributed memory architectures are primarily used. Classification structure of Memory Architecture is shown in figure 3.9.

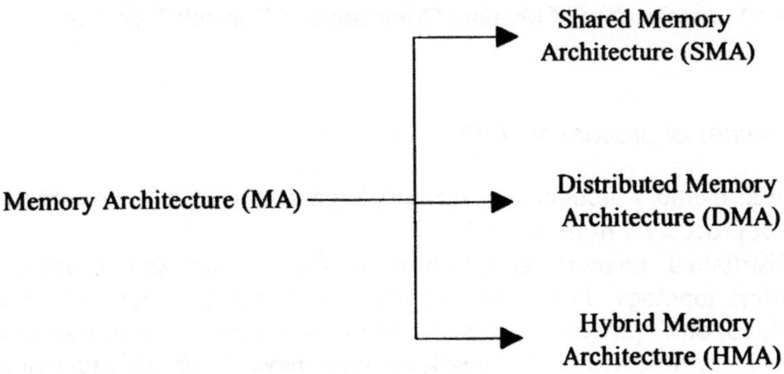

Figure 3.9 Classification of Memory Architecture

3.3.5 Memory Organisation / Network Topology (NT)

Processors can be connected in many different interconnection topologies. Based upon the interconnection topologies, shared memory architecture can be classified as global memory organisation and global-plus-local memory organisation. Distributed memory architectures can be classified as Tree, Mesh, Cube, and Pyramid architectures. Classification structure of Memory Organisation / Network Topology is shown in figure 3.10.

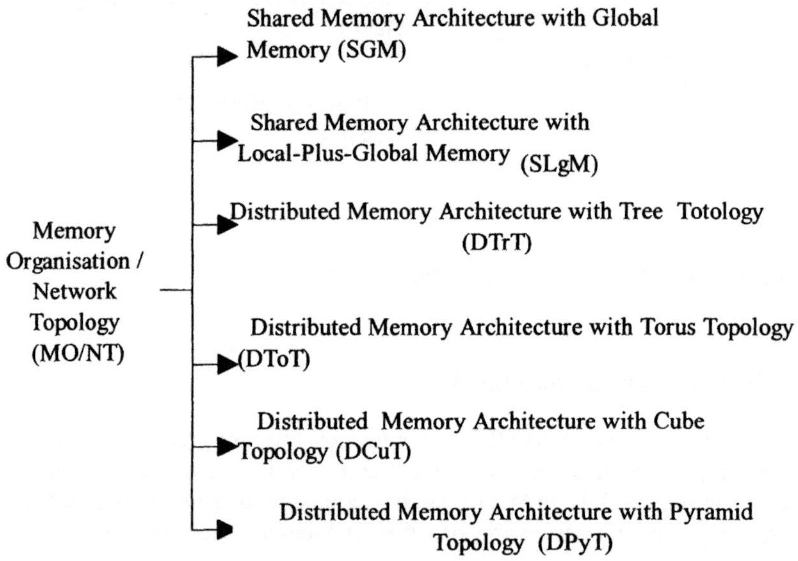

Figure 3.10 Classification of Memory Organisation / Network Topology

3.3.6 Number of processors (NP)

In a shared memory architecture, memory organisation can be constructed with any number of processors from one to N.

In distributed memory architecture number of processors are fixed for a symmetrical topology. For example symmetrical tree topology can be constructed with 3, 9, 15, or 27 processors. Symmetrical torus topology can be constructed with 5, 9, and 17, symmetrical cube topology may have 5 or 28 processors, whereas symmetrical pyramid topology can have 5, 9, or 21 processors. The number of processors used in different symmetrical network topologies is shown in figure 3.11.

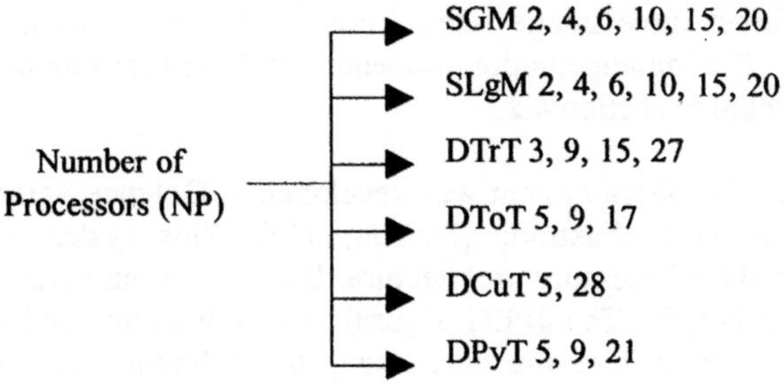

Figure 3.11 Classification of Number of Processors

3.4 Implementation of Plans on Parallel Computer Architectures

Some of the digital image compression Plans were implemented on parallel computers, while many others were simulated using the Network II.5 package. A brief introduction to the computers used for implementing some of the Plans is given in section 3.4.1. Simulation steps used for modelling and simulating other Plans are given in section 3.4.2.

3.4.1 Implementation of Digital Image Compression Plans on Parallel Computers

Plans for Non-Block Comparator Technique (NBCT) were implemented on parallel computers such as, Mercury system, Shiva system, and Param system. These computers are course grained and are described in the following sections.

Mercury system: The Mercury system was designed and developed by the Collaborative Information of Technology and Research Institute (CITRI), Melbourne [Bevinakoppa, 92]. It is a message passing parallel computer. Mercury system comprises four T800 transputers and sixteen T805 transputers and operates as back-end processor to a 386 based host processor. The JPEG algorithm was implemented on various number of processors on Mesh topology distributed memory architecture using the Plan,

Mercury Plans := P(NBCT, NIPC, BBIP, DMA, DToT, NP)

where,
NP = 1 | 2 | 3 | 5 | 9 | 17,
Image Size = 125 x 125 | 228 x 231 | 625 x 423.

This Plan was implemented on the Mercury system with NP = 1, 2, 3, 5, 9, and 17 for three different three image sizes. Thus a total of eighteen (6 x 3) Plans were implemented. The implementation procedure of these Plans on the Mercury system is explained in detail in section 4.2.

Shiva system: The Shiva system was developed at Defence Science and Technology Organisation (DSTO) Australia [Nelson, 92b]. This system is a multiprocessor designed with shared memory architecture. It can accommodate upto eighteen Intel's i860 processor boards. The JPEG algorithm was implemented on this global-plus-local memory architecture machine using the following Shiva Plans. Details of implementation of these nine Plans are given in section 4.3.

Shiva Plans := P(NBCT, NIPC, BBIP, SMA, SLgM, NP)

where,
NP = 1 | 2 | 3
Image Size = 125 x 125 | 228 x 231 | 625 x 423.

Param system: The Param system was designed and developed by the Centre for Development of Advanced Computing (C-DAC), India [Bhatkar, 94]. Param is a multiprocessor system that uses message passing, as well as shared memory parallel programming. Each node of the Param system contain four Transputers and one i860 processor. The JPEG algorithm was implemented on this hybrid architecture with various number of processor connected in Mesh topology. Plans used on this architecture may be represented by,

Param Plans := P(NBCT, NIPC, BBIP, DMA, DToT, NP)

where,
NP = 1 | 2 | 3 ,
Image Size = 125 x 125 | 228 x 231 | 625 x 423.

The Detail of these nine implementations are explained in section 4.4.

These three machine were used to implement a total of twenty six Plans. More than two hundred other Plans were simulated using the Network II.5 simulation package.

3.4.2 Simulation of Parallel Processing Plans for Image Compression

Simulation allows the assessment of potential performance before operating newly designed systems. It permits the comparison of various operating strategies of a system when the real system is not available. It allows measurement of parameters such as the time taken for various operations, utilisation of various components and system efficiency.

Network II.5 package was used to model the operation of various Plans for parallel processing of the JPEG algorithm. The execution times of these Plans were obtained

by running various simulations on the Network II.5 models. Simulation results are given in chapter 5. The procedure adopted for developing the models, and running the simulations is explained in this section.

Model building and simulation involves ten steps, viz. problem formulation, model building, data collection, model translation, model verification, model validation, experiment planning, experimentation, analysis of results, and documentation [Sharda, 95]. Each of these steps are described briefly in the following sections.

3.4.2.1 Problem Formulation: The first step in building a model is to make a clear statement of the problem. This includes stating the objectives of the modelling and simulation project. An accurate and precise problem statement is very essential for a systematic and smooth operation of the simulating process.

The main aim of our simulation work is to test the operation of image compression algorithms on different parallel computer architectures. To systematise the modelling and simulation work the idea of a Plan was introduced, in section 3.3. Each Plan was translated into a model using the Network II.5 package. Performance figures for these Plans were obtained by running simulations on these models.

3.4.2.2 Model Building: There can be many different types of models such as descriptive models, physical models, mathematical models, flow charts, schematics and computer based models. The computer based models was chosen for discrete event simulation.

The criteria used in selecting the best type of model are [Leigh, 83]:

- Design a model that could be used for different applications with some variations in the model [Reitman, 81].
- Simulation models may work for some specific real systems and may not work for other systems. Because, hidden critical assumptions may cause the model to diverge from reality. Therefore, it is a better idea to design a model for a specific problem with all the required parameters.
- The model output should be close to the expected output value. If not, the above steps need to be followed again from the beginning.

The various Plans, mentioned in section 3.3, were modelled for parallel processing of the JPEG algorithm. The following steps were followed for each of the models built.

3.4.2.3 Data Collection: Data for various parameters of the model e.g. that of a parallel computer system can be collected from various sources. Data, such as values of various parameters of the model, may either have a single value or a range of values. These values can be collected through literature survey and/or initial experimental results.

Data required for modelling the execution of Digital Image Compression techniques on parallel computers comprises data related to different compression techniques, and data required for modelling parallel computers. For the Non-Block Comparator technique data was collected from the existing JPEG algorithm described in the current literature. For the Block Comparator technique data was obtained from the experimental results, by running the block comparison step on Sparc II processor several times. From the experimental results it was found that parameters, such as execution time, followed Normal Distribution.

Distributed, shared, and hybrid memory architectures were used for modelling some of the Plans. The parameters required for modelling these architectures were taken from the manufacturer's specification for the three parallel machine (Mercury, Shiva, and Param) described in section 5.2.

3.4.2.4 Model Translation: Next step is to translate the model into a form which can be processed on a computer. Model translation can be done in two ways, either by using a suitable computer language, or by using a simulation package.

Simulation programs can be written in high level programming languages such as Fortran or Pascal. Some languages are designed specifically for simulation, such as SimSript, GPSS, Simula [Bratley, 83] [Naylor, 68]. A special simulation language such as Continuous System Simulation Language (CSSL) or Modified Integration Digital Analogue Simulator (MIDAS) is extremely easy to use [Stephenson, 71].

Some simulation packages target towards specific application areas are also available. For example Network II.5, COMNET II.5 and SIMLAN II can be used to model computer networks. Network II.5 simulation package was chosen for simulation of digital image compression techniques on parallel computers. Network II.5 is described in the following section.

Network II.5 simulation package: Network II.5 is a simulation package which takes in the description of the computer hardware and software and gives reports on measures such as hardware utilisation, software execution, times taken for message delivery, contention of memory, total system performance etc [CACI, 94]. It has an interactive graphical interface and can generate diagrams of the computer, storage devices, and transfer devices. Network II.5 models the interactions between all the devices in the system.

Network II.5 package has three main functions, viz., system description, system simulation and system analysis. In system description, all the devices or components are represented by graphical icons and their attributes can also be specified via a graphical interface. Simulation results can be analysed after running various experiments on the models. Network II.5 facilitates this analysis through animation, plot of device status and reports.

There are two categories of entities in Network II.5; hardware and software. These are described in following sections.

Hardware description on Network II.5: Hardware devices are modelled as one of the three building blocks available in Network II.5; these are, Processing Element (PE), Transfer Device (TD), and Storage Device (SD), depending on the function of the device being modelled.

The Processing Element building block can be used to simulate any intelligent hardware devices such as bus controller, display controller, and Central Processing Unit (CPU). A PE is characterised by parameters such as basic cycle time, message list size, I/O setup time, time slice, interrupt overhead, input controller, and instruction repertoire. The instruction repertoire is classified into four groups of instructions; processing instructions, read/write instructions, message instructions and semaphore instructions.

Transfer Devices are the links between Processing Elements and Storage Devices or between two Processing Elements. Transfer Devices are characterised by parameters such as transfer speed, transfer overhead, number of words, number of blocks and protocol definition. A message instruction is used to move the data between two PEs and read/write instructions are used to move the data between PE and storage device. Network II.5 automatically computes the actual time to send the data, and organises the data transfer according to the specified transfer protocol. The transfer protocol attribute may be set to model First Come First Serve (FCFS), collision, token ring, token bus, priority, aloha, and other protocols.

A Storage Device (SD) can be connected to more than one Processing Elements. Storage Devices are used to store data in files. The capacity of these is measured in bits. Read / write instructions are used to read the data from file and write the data to a specified file. If a specified file does not exist, Network II.5 gives a warning message at run time. Only a portion of the data can be accessed at any time. If the file structure is yet to be determined or is not significant to the simulation, files can be read from, or write to, a general storage (GS). The general storage keeps track of the number of bits stored.

Software description on Network II.5: In a Network II.5 based simulation, algorithm (software) is modelled in the form of software modules. Each module contains a list of PEs on which it may execute, a description of when it may run i.e. module preconditions, what it is to do when running i.e. instruction list, and which other modules to start upon completion i.e. successors. Each module goes through four cycles, these are;

1. Check preconditions
2. Once all preconditions are met queue up for PEs
3. When the PE becomes available execute instructions from its instruction list.
4. After the module has issued the last instruction in its instruction list, choose its successor modules, if any.

3.4.2.5 Model Verification: The purpose of the verification step is to ensure that the model behaves as planned and that it is a true representation of the system being modelled [Roberts, 83].

Network II.5 facilitates verification by providing graphical output for the hardware and software components of the model. The complete (hardware plus software) model can be verified by running animation. This enables visualisation of the modelled system in operation. The operation of each implementation can be seen visually and can be compared against the planned model. The animation operation can be performed as a step-by-step operation or as a continuous operation.

3.4.2.6 Model Validation: A model can be validated by proving that the model is a correct representation of the real system. There are various techniques for validation ie. mathematical technique, experimental, or statistical. The best way to validate a model is to compare the results of the simulation with results produced by the real system operating under the same conditions. If the compared result is within +/- 10% of the predicted value then the model is said to be validated.

In our research some of the Plans were implemented on real systems such as Mercury system, Shiva system, and Param system. The same Plans were simulated on Network II.5 package. Validation for our system was done by comparing the experimental results obtained on the real systems with the results obtained from the Network II.5 based modules.

3.4.2.7 Experiment Planning: To run a series of experiments we must plan for the values over which the variables would be varied; because only a finite number of experiments can be ran on each model. To draw a useful conclusion, for some system we may be able to plan the experiment before the simulation starts. For complex systems, the later experiments have to be planned based on experience from the initial experimental results.

For our simulation, various Plans were simulated. For each Plan the Number of Processors parameter was varied from low values to high values till the speedup started decreasing. From these series of experiments the speedup and scaleup performance measures were determined.

3.4.2.8 Experimentation: A total of ten Plans were modelled, and, on an average, thirty experiments were run for each Plan. Therefore, a total of three hundred experiments were run on the Network II.5 based models.

3.4.2.9 Analysis of Results: A vast volume of data is generated from the large number of experiments carried out. This data is plotted in a series of graphs to be able to study this data. These graphs are then analysed to derive conclusions.

3.4.2.10 Documentation: The output data produced by these experiments must be well documented. Documentation is essential for reuse and maintenance of the model.

In our system all the Plans which are represented in graphical objects were documented, and were stored on a computer disk using ABC flow chart. The experimental results obtained for all the Plans were maintained on disk using Excel spreadsheet. Speedup were calculated using the spreadsheet software and graphs of each Plan were drawn using Microsoft Draw.

3.5 Performance Measures

Parallel processing performance is measured in terms of speedup, scaleup and efficiency of the system. These terms are defined in the following sections.

Speedup of a parallel processor: Speedup for N processors is defined as the time taken by a single processor divided by the time taken by N processors [Kumar, 94]. Speedup (S_N) of parallel processor is given by,

$$S_N = T_1 / T_N \quad (3.1)$$

where,

S_N = Speedup for N processors,

T_1 = Time taken by a single processor,

T_N = Time taken by N processors.

Speedup curves can be of one of three types: superlinear, linear or sublinear as shown in figure 3.12.

Linear speedup is obtained when the improvement in performance is proportional to the number of processing elements in the system. Superlinear speedup is obtained when the speedup curve is above the linear curve. But, in most of the real implementations, linear or superlinear speedups cannot be obtained due to communication overhead. When the speedup curve lies below the linear curve sublinear speedup is obtained.

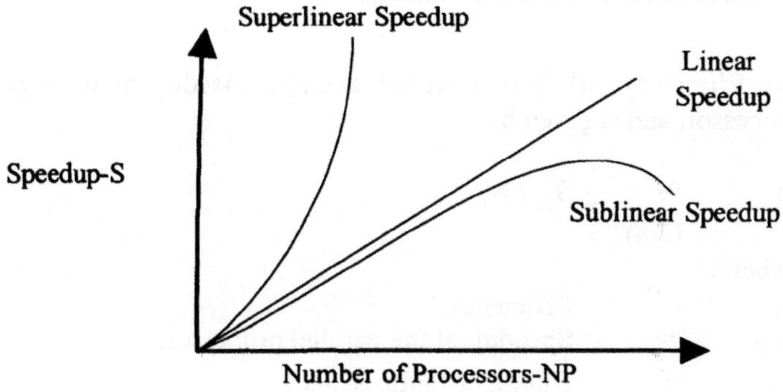

Figure 3.12 Speedup graph

Scaleup: For sublinear speedup, initially the speedup increases as the number of processors in the system increases, and at some point the speedup starts decreasing with further increase in the number of processors, as shown in figure 3.13. Scaleup of a parallel architecture is a function of the maximum number of processors at which the speedup starts decreasing. For brevity, scaleup is defined as the number of processors at the point of maximum speedup.

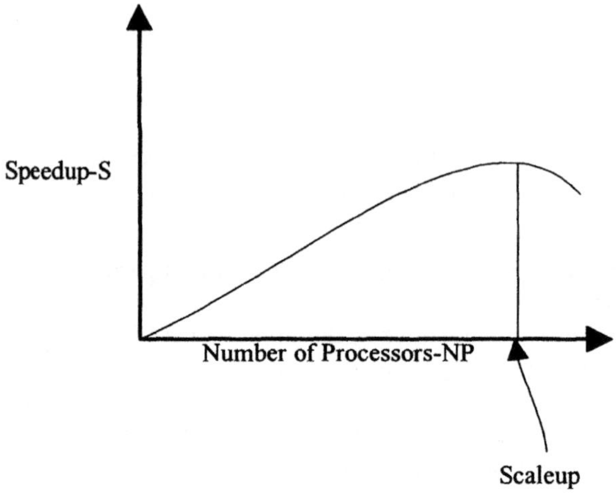

Figure 3.13 Speedup graph showing scaleup

Efficiency: Efficiency (η) is defined the average speedup of each processor in a parallel processor, and is given by,

$$\eta = S_N / N, \qquad (3.6)$$

where:

η = Efficiency,
S_N = Speedup of the parallel processor,
N = Number of processors.

3.6 Summary

Digital image compression can be implemented on parallel computers to speedup the operation. This chapter explained the different types of parallel computers and parallel programming languages. Various Plans for digital image compression on parallel computers were given. Some of these Plans were implemented on available parallel computers such as, Mercury, Shiva and Param systems. Some other Plans were simulated using Network II.5 simulation package.

Performance measures such as speedup, scaleup, and efficiency for parallel architectures were defined. These performance measures will be used in chapter 5 to evaluate the performance of various Plans for the parallel processing of JPEG algorithm.

4 IMPLEMENTATION OF THE JPEG ALGORITHM ON THREE PARALLEL COMPUTERS

4.1 Introduction

This chapter explains the hardware architecture and implementation of the JPEG algorithm on the parallel computer systems; Mercury, Shiva and Param.

The Mercury system uses a message passing distributed memory architecture. The Shiva system is a shared memory parallel architecture and has a torus interconnection topology. The Param system has a hybrid architecture, with distributed as well as shared memory.

The JPEG algorithm was implemented on the three above mentioned parallel computers with different image sizes on various sized networks. This chapter describes the implementation procedure and the experimental results obtained on each of these systems.

This chapter describes the implementation of the JPEG algorithm on three parallel computers namely; Mercury system, Shiva system and Param system.

The Mercury system was designed and developed by the Collaborative Information Technology and Research Institute (CITRI), Melbourne. It is a distributed memory, message passing parallel computer. The Mercury system comprises of twenty transputers and is used as a back-end processor to a personal computer.

The Shiva system was developed at the Defence Science and Technology Organisation (DSTO), Australia. It uses shared memory architecture and is used as a back-end system to a Sparc station. Shiva comprises of one Master unit and multiple Slave units. Slave units may be of any type, viz. Intel i860 based processors or a special purpose processor.

The Param system was designed and developed by the Centre for Development of Advanced Computing (C-DAC), India. It is a multiprocessor system that uses message passing as well as shared memory parallel programming. Param systems are available in three series, viz. Param 8000, Param 8600 and Param 9000. The Param

8600 is based on Intel's i860 processor. The i860 based Param 8600 was selected for this research.

The JPEG algorithm was implemented on the above mentioned three systems. Section 4.2 describes the hardware architecture, operating system and implementation of the JPEG algorithm on the Mercury system. Section 4.3 describes the hardware, parallel programming environment and implementation of the JPEG algorithm on the Shiva system. Section 4.4 describes the hardware architecture of Param 8600, Paras parallel programming environment and implementation of the JPEG algorithm on the Param system. Performance comparison of these three parallel computers is given in section 4.5.

4.2 Implementation of the JPEG Algorithm on the Mercury System

The JPEG algorithm was implemented on the transputer based Mercury system using the Helios operating system.

Section 4.2.1 describes the Mercury system and its parallel programming environment. Section 4.2.2 describes implementation of the JPEG algorithm on Mercury system. The JPEG algorithm was implemented on various sized network topologies and with different image sizes. Experimental results are given in section 4.2.3.

4.2.1 Mercury System Architecture

This section describes the Mercury system. Mercury system is a multi-user scalable Multiple Instruction Multiple Data (MIMD) parallel computer, with primary aggregate memory of 100 MByte, and twenty processors. It consists of four T800 transputers and sixteen T805 transputers.

4.2.1.1 Hardware Architecture: A 80386 based PC was used as the front-end to the transputer based Mercury system. It consists of twenty Transputers, four of which are T800s and sixteen transputers are T805s. Four transputers are connected in a ring topology and named as T1, T2, T3 and T4, as shown in figure 4.1. Sixteen transputers are connected in a torus topology, and named as N1, N2, N3 ... N16.

The INMOS transputer family is a range of system components each of which combines processing, memory and serial interconnection interfaces in a single VLSI chip, with a design based on the Reduced Instruction Set Computer (RISC) technology [INMOS, 89]. The first member of the family, the T414, a 32-bit transputer, was introduced in September 1985. Transputer architecture has inherent concurrency that can be applied to a wide variety of applications such as simulation, robot control, image synthesis, and digital signal processing. These numerically intensive applications can exploit large arrays of transputers in a single system. The following series of transputers are available; T800, T805, T9000 and Alpha. The overall performance of a transputer based system is dependent on the number of transputers, the speed of inter-transputer communication and the floating point performance of each transputer. The following section describes the characteristics of the T805 transputer.

T805 transputer architecture: The T805 has a floating Point Unit (FPU), a CPU, 4 KBytes of local memory, four communication links and a timer as shown in figure 4.2. INMOS T805 provides a peak computing power of 3.5 MFLOPS at 20 MHz. The Transputer has an internal memory of 4KByte, which is too small to run most applications. Thus external transputer Modules (TRAMS), which consist of 32 KBytes - 2 MBytes of memory are often used. Communication links are Direct Memory Access (DMA) based bi-directional links that can be used to connect many transputers in a multiprocessor system [Mitchell, 90]. The peak communication speed between two T800 or T805 transputers is 20 Mbits/sec.

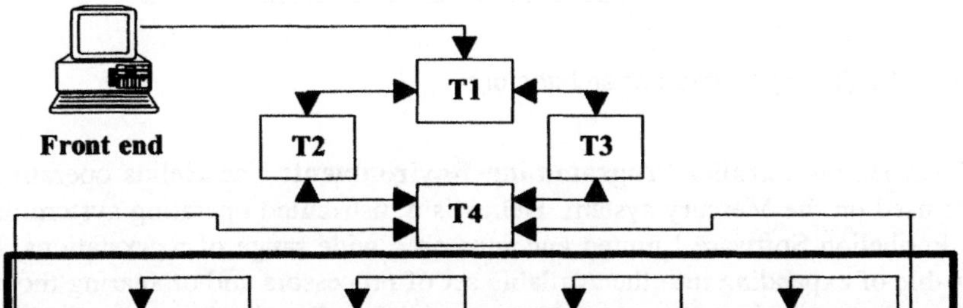

Figure 4.1 Interconnection topology of the Mercury system

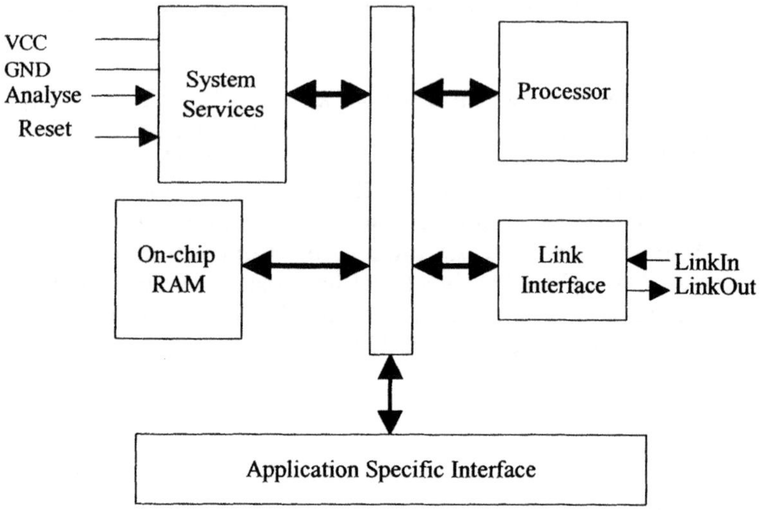

Figure 4.2 (T805) Transputer architecture

4.2.1.2 Helios Parallel Programming Environment: The Helios operating system was used on the Mercury system. Helios is a distributed operating system developed by Perihelion Software Limited and runs on a wide range of workstations. Helios is capable of expanding into the available set of processors and of sharing the workload among them. Such processor clusters may themselves be interconnected in a local network to allow the sharing of data and expensive devices such as high capacity discs and laser printers [Ian, 92] [Hemery, 91].

Helios provides four levels of communication between processors. The lowest level is used by the nucleus: PutMsg() and GetMsg() are the Message Passing Primitives (MPP) that provide the basis of all Helios communication. The level above this is provided by the system library functions Read() and Write(). These calls operate on streams and have time outs associated with each requested transaction. At the next level is the, Portable Operating Set of Instruction Codes (POSIC), read() and write() functions. Calls to the POSIC functions are based on POSIC file descriptors. The highest level of communication is at the language level, which depends upon the programming language used.

The POSIC communication routines is used most often because of the following factors.

1. This communication mechanism assures some degree of reliability and fault tolerance. That is, it provides error detection and recovery from failure and there is a guarantee that messages will arrive at their respective destinations.
2. It offers greater functionality than lower level libraries.
3. Most importantly, POSIC library can also be used on other parallel architectures, thereby giving portable code.

MPP as well as POSIC communication routines were used to compare their effect on the execution times. In these two levels of communication, inter-processor message transmission times over transputer links are characterised by a relationship of the form:

$$T_{total} = t_{overhead} + t_{init} + N.t_{tx}$$

$$= 126.120129 + 0.562684.N \quad \text{for MPP}$$

$$= 1461.421142 + 0.567642.N \quad \text{for POSIC}$$

where,

T_{total} = message transmission time,

$t_{overhead}$ = loop overhead on each test iteration,

t_{init} = message initialisation time,

N = number of bytes in a message,

t_{tx} = transmission time for one byte.

Some of the most useful facilities in the POSIC communication protocols are:

- open() is used to open a stream to a named file or server and it returns the file descriptor that can be used by other routines.
- link() is used to create a symbolic link, in other words an entry in the naming tree that actually refers to some other object elsewhere in the naming tree.
- close() terminates a stream connection to a file or server that was produced by open().
- read() attempts to obtain data from an open file or server. The read statement is written as:
  read(file descriptor, address of the buffer, length of the buffer).
- write() attempts to send data to an open file or server. The write statement is written as:
  write(file descriptor, address of the buffer, length of the buffer).

4.2.1.3 Component Distribution Language (CDL): CDL is the language that facilitates parallel programming under Helios. The purpose of CDL is to provide a high level approach to parallel programming. It allows Helios to take care of the actual distribution of the program components over the available physical resources.

An example of CDL script is shown below. In CDL script the sentence followed by # is a comment. The script consists of two parts, viz. component declaration(s) and task force definition.

The purpose of the component declaration is to specify relevant details of the hardware component to the Helios operating system. The task force definition is a specification of the network topology, and is used by the Helios operating system.

# This is a CDL script example	Comment
component master {memory 20000; puid /Cluster/T1; }	Component declaration
master | slave1 <> slave2	Task force definition

The component declaration part describes the requirements of particular components in the task force. A component can be declared in terms of memory size, path of the processor location, and name of the processor, as shown in the above example.

The task force definition part describes the interaction of the task force with the components. The CDL language defines four parallel constructors, i.e. | pipeline constructor, <> bi-directional constructor, ^^ parallel constructor, and ||| interleave constructor.

4.2.1.4 Parallel Programming Languages: Transputer based systems support OCCAM, C, and C++ programming languages [Ungerer, 91]. The transputer was designed to execute the OCCAM parallel programming model efficiently. OCCAM programs can be operated on four independent channels in parallel [Pountain, 87].

4.2.2 Implementation of the JPEG Algorithm on the Mercury System

The JPEG algorithm was implemented on the Mercury system using the C programming language under the Helios operating system. It was implemented on various number of processors on Mesh topology distributed memory architecture using the Plan,

Mercury Plans := P(NBCT, NIPC, BBIP, DMA, DToT, NP)

where,
P is Plan for implementation,
NBCT is the Non-Block Comparator Technique used for image processing,
NIPC is Block Dependency with Non-Inter-Processor Communication,
BBIP is Block Based Image Partitioning method,
DMA is Distributed Memory Architecture,
DToT is Distributed Memory Architecture with Torus Topology,
NP is Number of Processors = 1 | 2 | 3 | 5 | 9 | 13 | 17.
Image Size = 125 x 125 | 228 x 231 | 625 x 423.

This Plan was implemented on the Mercury system with NP = 1, 2, 3, 5, 9, and 17 for three different three image sizes. Thus a total of eighteen (6 x 3) Plans were implemented. This section explains the implementation procedure of parallel JPEG implementation on the Mercury system.

The implementation procedure used on the Mercury system is shown in figure 4.3 as a flow diagram [Sharda, 93]. The source image is initially stored on the host processor T1. The Image is transferred to the T4 node processor through the T3 node processor. The node processors (N1 to N16) wait for components of the source image to be down loaded. The T4 processor partitions the image into two parts and sends each part along with the required header information to node processors N1 and N4. For image partitioning a block of 8x8 samples is used as an atomic component. The N1 node processor further divides the image into eight parts and sends seven of these parts to node processors N13, N5 and N2. One sixteenth of the image is retained on the N1 node processor. This distribution of image parts continues as shown in the path graph given in figure 4.4. Thus, each node processor has 1/16th of full image. The Image distribution is followed by performing encoding on each processor in parallel. The encoded image is composed using a reverse procedure with respect to the procedure used for image distribution.

The above described procedure was used for distributing the image onto seventeen processors. Experiments were carried out to determine the execution time on fewer processors as well. For each experiment the image was distributed as evenly as possible on the set of processors being used.

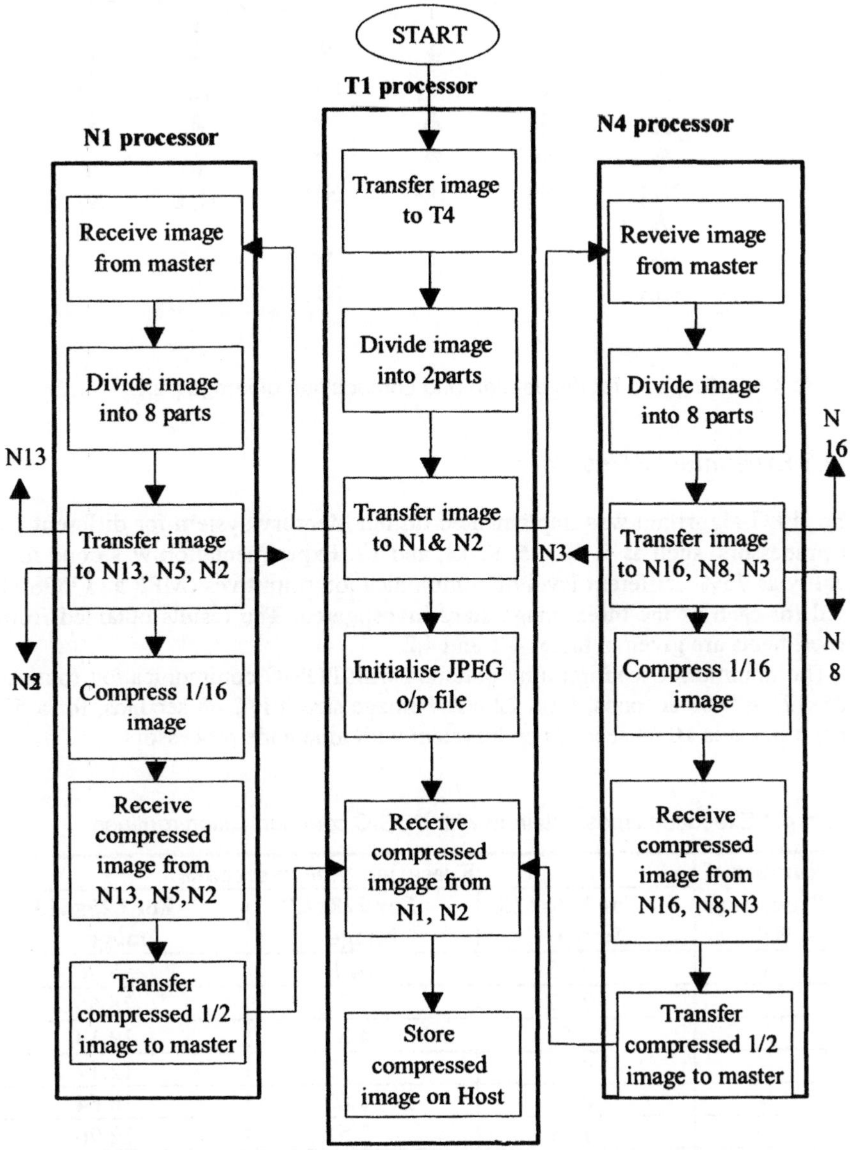

Figure 4.3 Flow diagram of implementation procedure on the Mercury system

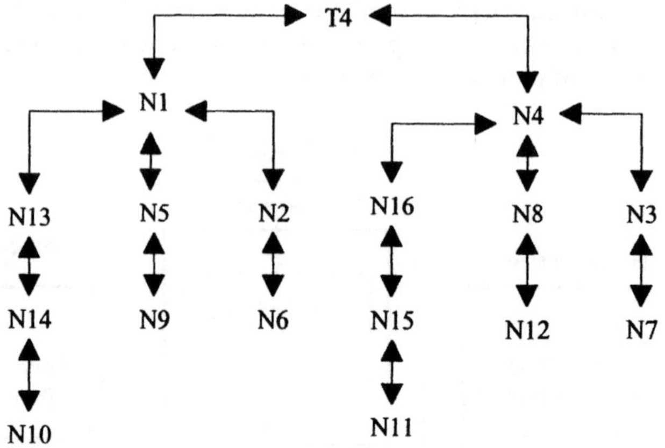

Figure 4.4 Path graph for distribution and composition of image parts

4.2.3 Experimental Results

The JPEG algorithm was implemented on the Mercury system for different numbers of processors, such as 1, 2, 3, 5, 9, 13, and 17. Experimentation was done for output quality as 75%. Different level of communication primitives (MPP and POSIC) were used for each of the three image sizes investigated. The results obtained from these experiments are given in tables 4.1 and 4.2.

The minimum execution time obtained with POSIC communication routines for a 125x125 is 0.89 seconds, for a 228x231 image size it is 2.88 seconds, for a 625x423 image size it is 10.64 seconds on a system with nine node processors.

Table 4.1 Execution times obtained with POSIC communication routines

Number of Processors - NP	**Execution time in seconds**		
	For 125x125 image	**For 228x231 image**	**For 625x423 image**
1	4.35	14.14	55.86
2	2.27	7.33	28.65
3	1.53	4.89	19.13
5	0.97	3.13	12.17
9	**0.89**	**2.88**	**10.64**
13	1.19	3.52	12.96
17	2.26	6.73	19.33

Table 4.2 Execution times obtained with MPP communication routines

Number of Processors - NP	Execution time in seconds		
	For 125x125 image	For 228x231 image	For 625x423 image
1	4.35	14.14	55.86
2	2.25	7.25	28.35
3	1.51	4.86	18.87
5	0.94	3.04	11.86
9	**0.69**	**2.09**	**7.66**
13	0.83	2.38	8.85
17	1.22	2.97	10.27

4.3 Implementation of the JPEG Algorithm on the Shiva System

Around 1990, the Defence Science and Technology Organisation (DSTO), Australia undertook a research project on a heterogeneous, parallel multiprocessor named as the Shiva[1] system [Yakovleff, 91]. The aim of the project was to enhance overall performance by incorporating several data paths in the architecture in order to achieve efficient and balanced processor utilisation. It incorporated performance enhancing techniques, such as multiplicity, heterogeneity, and reconfigurability.

4.3.1 Shiva System Architecture

Shiva is a heterogeneous, shared memory, multiprocessor parallel architecture. It is designed to exploit the I/O and operating system features of existing computers. It is intended to supply its parent system with enhanced performance over a wide range of applications without the need of any special parallel programming on the part of the user [Anderson, 90]. It can act as a Multiple Instruction Multiple Data (MIMD) computer, or as a pseudo Single Instruction Multiple Data (SIMD) computer.

4.3.1.1 Hardware Architecture: The Shiva system has eighteen processor units. The processor unit has Intel i80860 processor. The first processor unit is called the Master unit and the others are called the Slave units. Each processor unit has its own processing element (PE) and 16 MBytes of local memory [Karnak, 92a]. Local memory can be accessed either directly by the resident processor via a hotline or through a bus to which each processor has access as shown in figure 4.5. The Master unit contains the following elements:

- Co-ordinator,
- Memory unit,
- SBus interface,
- Subsystems such as Bootstrap EPROM, real-time clock, serial interface, registers,
- Bus arbitor.

[1]Shiva is the Hindu God of creation.

The control signals to and from the i860 are handled by a co-ordinator which includes address/parameter FIFOs to make use of the processor's pipelining capabilities. The co-ordinator maps requests from the i860 to the various devices (local memory, SBus or subsystem) or to the arbitrator if any of the other memory units is to be accessed.

The Slave units contain devices, such as co-ordinator, memory unit, data pipeline. All Slave units need not be of the same type, implying the possibility of a heterogenous architecture as shown in figure 4.6. Slaves may be special purpose boards such as a Neural Accelerator Board (NAB) and Parallel Transformation board (ParaT) and Intel i860 based processors [Anderson, 92] [Nelson, 92b]. NAB is used for graphical simulation in real-time [Nelson, 93b]. ParaT system is used for performing stereoscopic terrain visualisation application [Yakovleff, 94] [Nelson, 92b].

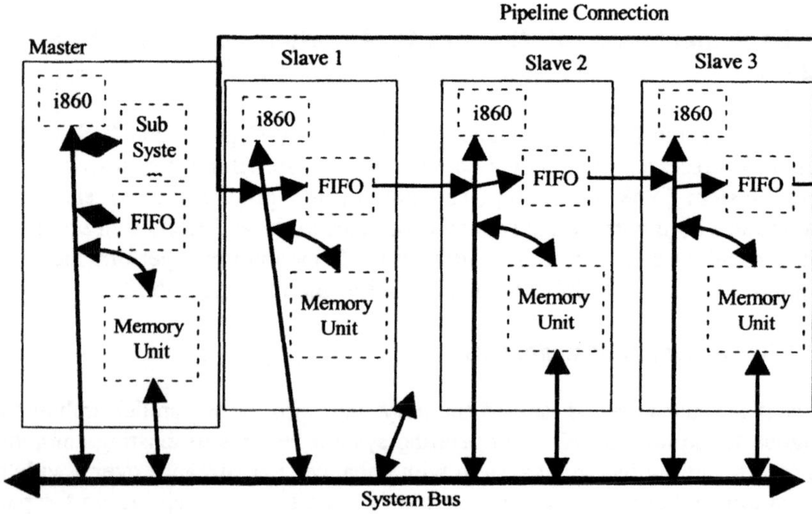

Figure 4.5 Master and Slave units and data paths

Intel i80860 architecture: The Intel i860 was designed for numerical and vector intensive applications [Margulis, 90]. Many of the design principles used have been adopted from supercomputer technology enabling the i860 to deliver a peak arithmetic performance of 80 MFLOPS (single precision) and 60 MFLOPS (double precision) in conjunction with a peak integer performance of 40 MIPS. In particular, its high throughput is achieved from a combination of RISC design techniques, pipelined processing units, wide data paths, and large on-chip caches. On a single chip, the architecture supports the following facilities, as shown in figure 4.7:

- Integer operations,
- Floating point operations,
- Graphical operations,
- Memory-Management support,
- Data Cache and Instruction Cache.

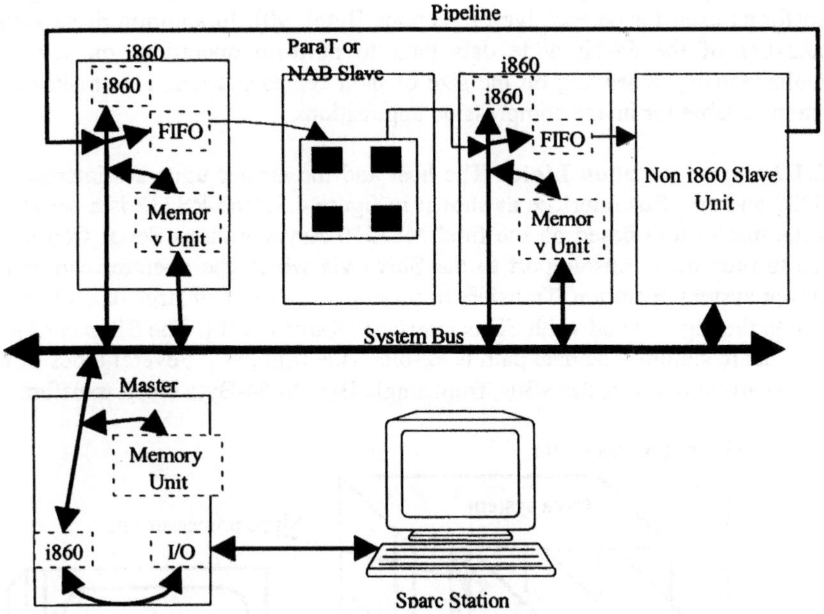

Figure 4.6 Shiva system with ParaT or NAB Slave units

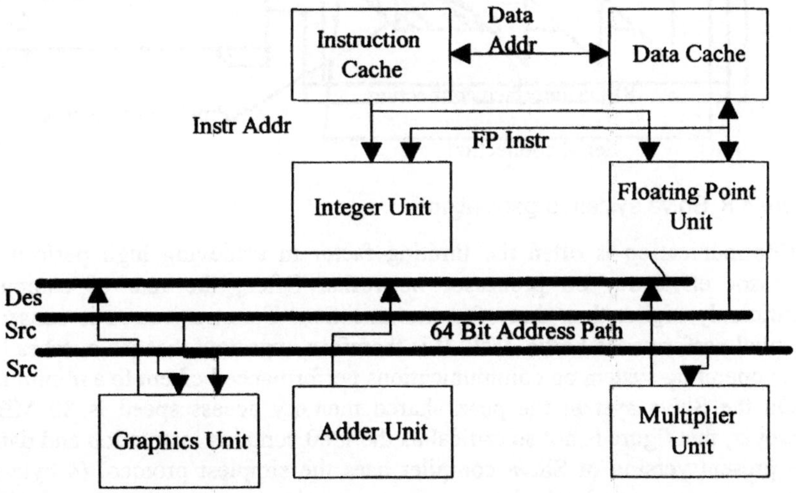

Figure 4.7 Intel i860 processor architecture

The i860 microprocessor has a number of instructions designed to perform operations specific to graphics, such as scan-line rendering, Z-buffering and 4x4 transforms used for perspective projections [Intel, 90]. In addition these instructions make use of the 64-bit wide data path to perform operations on several pixels simultaneously, depending on the size of their representation. This makes the Shiva system suitable for image compression applications.

4.3.1.2 Communication Links: The host and the master units are interfaced via an RS232 and an SBus interface as shown in figure 4.8. The RS232 is a serial interface on the master unit based on the Intel M82510 Asynchronous Serial Controller. It is used to provide a console port to the Shiva via which the operator can control and monitor system operation. Transfers of program binaries and large data blocks is done through the higher bandwidth SBus interface [Karnak, 92b]. The SBus card is located in the Sparc station. The data path is 32-bits wide [Sun, 90]. Several types of transfers can be carried out over the SBus, from single Byte to 64-Byte block transfers.

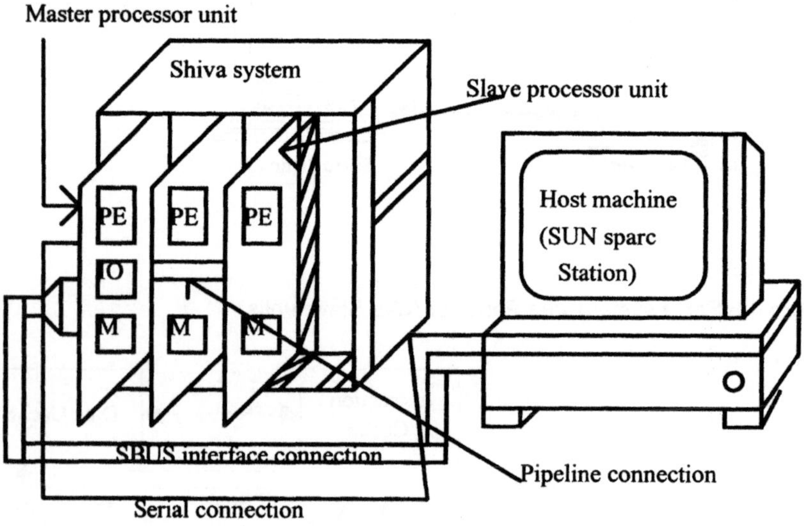

Figure 4.8 Shiva system organisation

Communication is often the limiting factor in achieving high performance and processor efficiency on processor networks. Often, the rate of computation is significantly higher than that of communication. Communication is consequently a potential performance bottle-neck. It is therefore important that the overhead imposed by an operating system on communications performance be kept to a minimum.

On the Shiva system the peak shared memory access speed is 80 MBytes/sec. However, this figure is not so critical as the i860 contains instruction and data caches. The present version of Shiva compiler uses the simplest protocol (4-byte read and write operations only) which can achieve roughly a contention free 5 MBytes/sec transfer rate over the SBus. Table 4.3 shows the data transmission times (t_{total}) and the rates with respect to message sizes (N) [Bevinakoppa, 94b], where,

N = Message size (bytes)
T_{total} = transmission time (microseconds)
R = Rate of transmission (KBytes/sec)

The rate of transmission with respect to the message size is plotted in a graph shown in figure 4.9. From this graph, it can be seen that transmission of a short message is inefficient. It is more efficient to transmit a single large message than a number of small ones. Transmission rate can be increased by implementing the burst ɔde transfer.

Table 4.3 Total time (T_{total}) and transmission rate (R) on the SBus interface for various message sizes

N (bytes)	T_{total} (microsec)	R (KBytes/sec)
1	1.886	530
4	4.78	836
16	12.4	1287
64	32.4	1970
256	96.6	2648
1024	296	3450
4096	947.76	4322
16384	3373	4856
65535	13133	4990

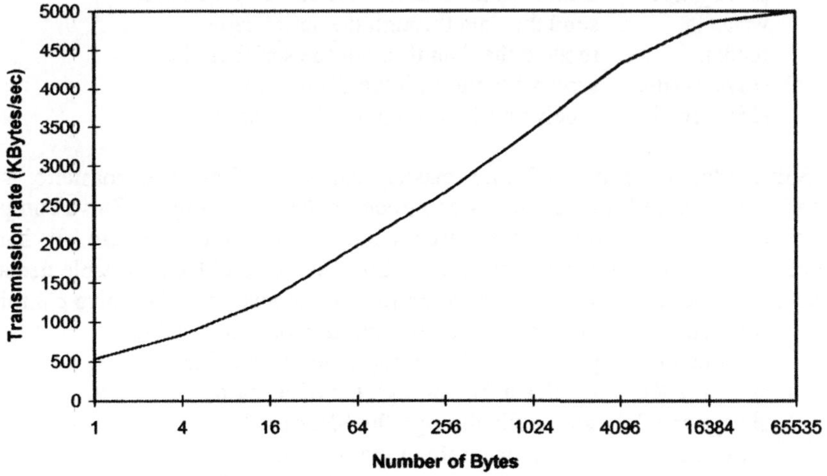

Figure 4.9 Rates of data transfer with respect to message size

One of the more novel aspects of the Shiva architecture is the inter-slave data pipeline. The pipeline and the shared bus, provide two mechanisms for inter-processor communication. Unlike the bus, the data pipeline is contention free. That is, all of the

Slave units can write to their data pipeline simultaneously. The pipeline is 64-bits wide and can support a write (and a read) every 4 clock cycles. This implies a peak bandwidth of 80MBytes/sec., which is the same as the peak memory and bus bandwidth. There are two modes of access for the pipeline: blocking and non-blocking. With a blocking access the requesting processor will be suspended if it attempts to read from an empty FIFO buffer or write to a full FIFO buffer. A non-blocking access will not suspend on a read from an empty buffer or a write to a full buffer. An attempt to write to a full buffer will result in the write data being lost and an attempt to read from an empty buffer will result in undefined data being returned. It is upto the controlling software to determine when it is appropriate to perform blocking or non-blocking pipeline operations.

4.3.1.3 Shiva Programming Environment: Most of the users are not interested in having to examine and modify their programs to take advantage of architectural features. Rather, users would prefer that the compiler and system software make the best use of machine's features. DSTO developed a pre-processor named Shiva compiler which can manipulate programs, so that parallelism may be extracted and control statements inserted to take advantage of Shiva's parallel processing features [Maurer, 88] [Nelson, 93a].

Programs on the Shiva system were divided into host files and Shiva files. Host files were compiled using a GNU C Compiler, that includes communication routines [Burns, 89]. Some of the communication routines are:

shiva_open:	which opens the connection between host and Shiva system ie. initialises SBus card and serial cable.
shiva_start:	initialises the Master and Slave processors.
write():	send the data through the serial cable.
read():	receive the data through the serial cable.
shiva_write:	send a file through the SBus card.
shiva_read:	receive a file through the SBus card.

Shiva files ware divided into master and slave files, e.g. master.c, slave1.c, slave2.c etc. These files were cross-compiled on the host using the Shiva compiler and downloaded to the Master processor using a shiva.out program. It has many functionalities such as communication routines, shared and local variable declarations, header files. Some of the communication routines available in the Shiva compiler are:

ser_gets():	get the data serially through the serial cable.
ser_putc():	put a character through the serial cable.
sbus_read():	receive a file through the SBus card
sbus_write():	send a file through the SBus card.
sem_send():	send semaphore to the Slaves.
sem_wait():	wait for the semaphores from the Slaves.

4.3.2 Implementation of the JPEG Algorithm on the Shiva System

Depending on the application, the programming structure can operate as a pure message-passing system with distributed memory, or a hybrid system comprising distributed clusters of processors with shared memory [Yakovleff,91]. A Shiva system based on three processors was used for experimentation. The JPEG algorithm was implemented on the global-plus-local memory architecture machine using the following Shiva Plans,

Shiva Plans := P(NBCT, NIPC, BBIP, SMA, SLgM, NP)

where,
P is Plan for implementation,
NBCT is the Non-Block Comparator Technique used for image processing,
NIPC is Block Dependency with Non-Inter-Processor Communication,
BBIP is Block Based Image Partitioning method,
SMA is Shared Memory Architecture,
SLgM is Shared Memory Architecture with Local-plus-global Memory topology,
NP is Number of Processors = 1 | 2 | 3,
Image Size = 125 x 125 | 228 x 231 | 625 x 423.

This section explains the implementation procedure of parallel JPEG implementation on the Shiva system.

The implementation procedure used on the Shiva system is shown in figure 4.10 as a flow diagram. In this flow diagram light boxes representing tasks are shown within dark boxes representing processors on which the tasks are executed. After initialisation, the Master processor waits for the source image to be sent by the front end. When the Master receives the image data from the host it sends a start signal to the Slaves indicating that the image is ready for compression. The Master and the Slaves processors start DCT, quantisation and encoding (compression) steps and store the compressed data in shared memory. The Master processor waits for the Slaves to finish compression, then it sends compressed image data to the host machine through the SBus. The host combines the data received from all processors and stores into the JPEG output file.

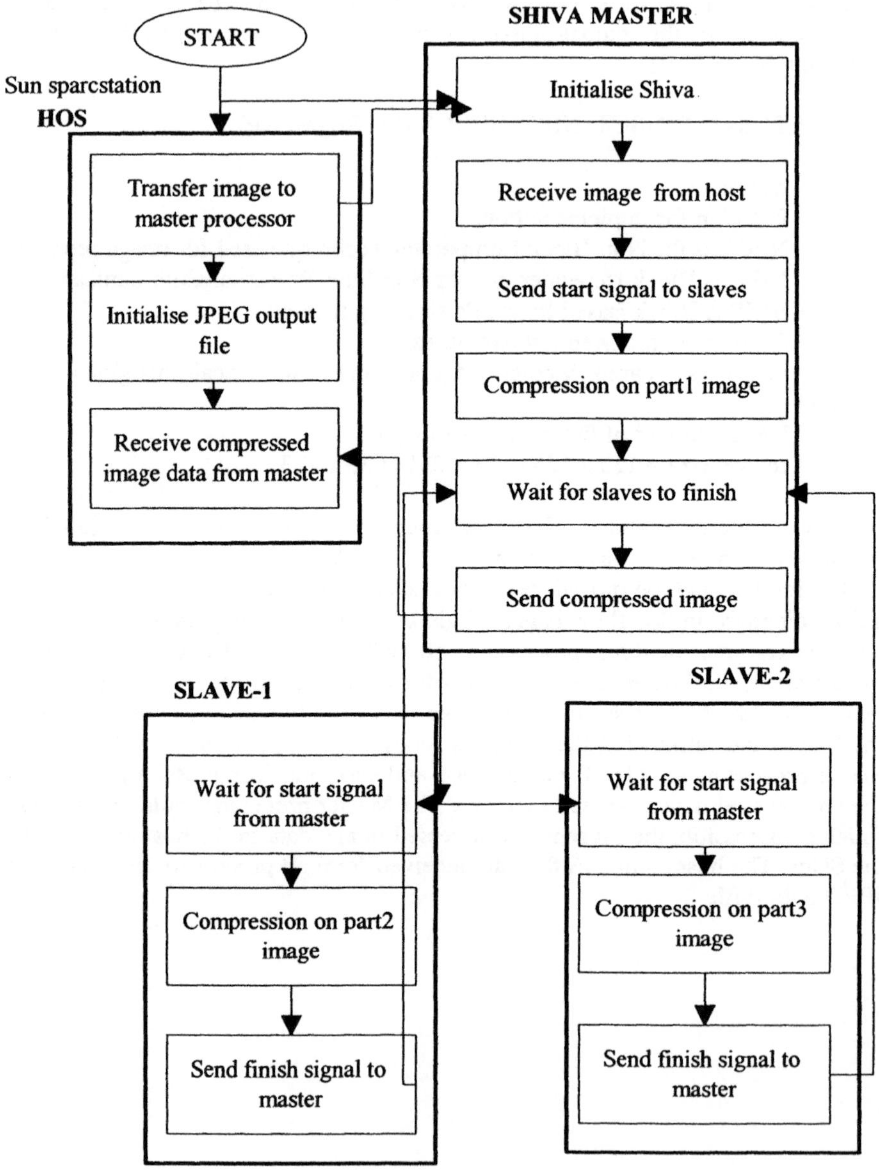

Figure 4.10 Implementation procedure of the JPEG algorithm on the three processor Shiva system

4.3.3 Experimental Results

The JPEG algorithm was implemented on the Shiva system on three processors for different image sizes and the results obtained on these are given in table 4.4. The results for four to seventeen processors were estimated by using the Gantt chart [Lewis, 92]. The procedure used for estimating the execution time obtained on three processors is as follows [Bevinakoppa, 94a].

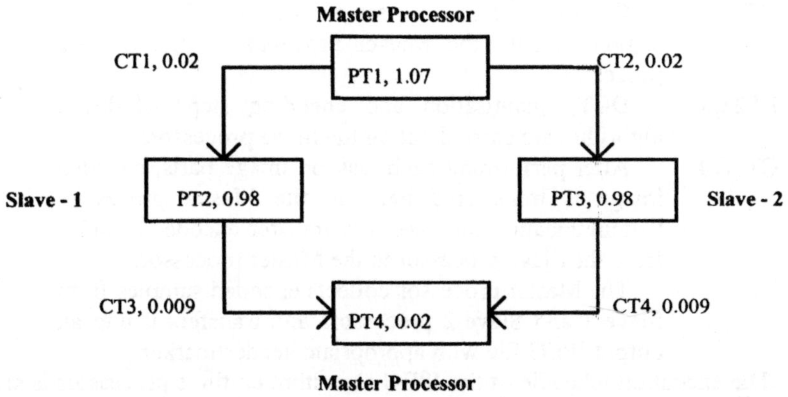

Figure 4.11 Task graph for three processors

The task graph for three processors is shown in figure 4.11. The JPEG algorithm is divided into four tasks as shown in the task graph of figure 4.11. In this task graph each processor represents a processing task. Inside each processor a task is represented as the ordered pair PTn, t(PTn) where PTn is the processing task number and t(PTn) is the processing time of that task. Arcs joining the processors represent communication tasks. A communication task is represented by the ordered pair CTn, t(CTn) where CTn is the communication task number and t(CTn) is its communication time. All times are in seconds (sec.). The processor on which a task is executed is specified adjacent to the task processor in the task graph. The values given below were obtained by using experimental results obtained on single i860 processor.

In this example 228 x 231 image was taken. Functions performed by the processing and communication tasks are as follows:

Task	Function	Execution time in sec.
PT1	Source image is divided on Master processor into three parts. Each part of the image contains 228 x 77 x 3 samples. First part of the image is transferred from the Master processor to Slave-1 processor. The Master processor performs DCT, quantisation and encoding steps on third part of the source image.	1.07
CT1	Communication time for transferring 228 x 77 x 3 samples from the Master processor to the Slave-1 processor.	0.02
CT2	Communication time for transferring 228 x 77 x 3 samples from the Master processor to the Slave-2 processor.	0.02
PT2&3	DCT, quantisation and encoding steps of JPEG algorithm are carried out on the Slave processors.	0.98
CT3&4	After performing each task on image parts, encoded image is transferred back to the Master processor. Communication time taken to transfer encoded samples from the Slave processor to the Master processor.	0.009
PT4	The Master processor collects encoded samples from Slave-1 and Slave-2 processors and transfers it into an output JPEG file with appropriate header/marker.	0.02

The execution schedule of the JPEG algorithm on three processors is shown in the form of Gantt chart in figure 4.12. A Gantt chart essentially shows the scheduling of various tasks on the time axis.

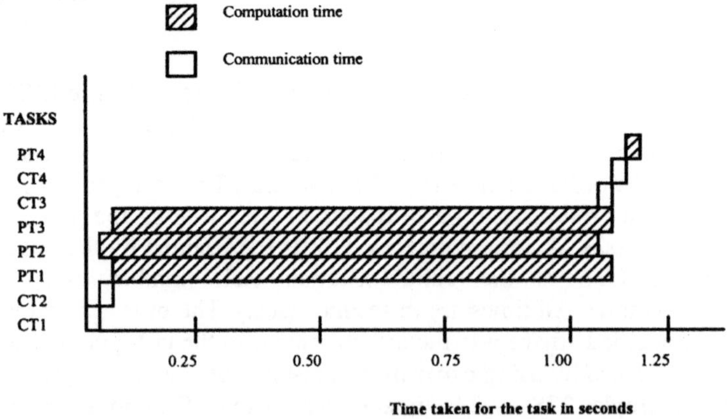

Figure 4.12 Gantt chart of JPEG algorithm on a three transputer network

From figure 4.12 it can be seen that the total time taken by the three Nodes

$$T3 = t(CT1) + t(CT2) + t(PT1) + t(CT3) + t(CT4) + t(PT4)$$
$$= 0.02 + 0.02 + 1.07 + 0.009 + 0.009 + 0.02$$
$$= 1.148 \cong 1.15 \text{ sec.}$$

The estimated execution time on a three node processors is nearly equal to 1.15 sec. which is the same as the execution time obtained from the actual implementation. The execution times for four to seventeen processors were estimated from the computation time and communication time obtained on a single node processor. These results are tabulated in table 4.4.

The minimum execution times obtained for a 125x125 image size is 0.12 seconds, on a system with thirteen node processors. Whereas, for 228x231 and 625x423 image sizes are 0.53 seconds and 1.32 seconds respectively, on a system with nine node processors.

Table 4.4 Execution times of the JPEG algorithm on the Shiva system

Number of Processors - NP	Execution times in seconds		
	For 125x125 image	For 228x231 image	For 625x423 image
1	0.6	2.63	7.02
2	0.32	1.34	3.55
3	0.28	1.15	2.41
5	0.15	0.60	1.52
7	0.13	0.55	1.39
9	0.13	**0.53**	**1.32**
13	**0.12**	0.60	1.60
17	0.20	0.91	2.58

4.4 Implementation of the JPEG Algorithm on the Param System

The Param[2] system is a distributed memory, message passing parallel computer developed by the Centre for Development of Advanced Computing (C-DAC), India [Bhatkar, 91], [Tulshibagwale, 94]. Param works as a back-end compute engine to hosts such as PCs, SUN workstations, MicroVAX machines and U6000 Unix machines. The Param architecture can accommodate heterogeneous nodes such as disk I/O nodes, graphics nodes, transputer nodes, vector nodes based on the Intel i860, and DSP nodes based on the Zoran 34325. Param is available in three series of configurations. The Param 8000 series is a replicate scalar processor machine based on the T805 transputer, which can be configured with 16, 32, 64, 128 or 256 nodes, and if required with more than 1024 nodes. The Param 8600 series is equipped with vector processing capabilities and is based on the Intel i860 processor. The Param 9000 is based on the SPARC II processing nodes. Since most of the digital image compression techniques are DCT based, the i860 based Param 8600 was chosen for this study.

Section 4.4.1 describes the Param system architecture. The JPEG algorithm was implemented on the Param system using the Paras parallel programming environment. Section 4.4.2 describes the implementation procedure employed on the

[2] Param is an acronym for PARAllel Machine, and in Sanskrit it means Supreme.

Param system. Experimental results were obtained on a three node Param system. Section 6.4 gives the results obtained for different image sizes on various sized networks.

4.4.1 Param System Architecture

This section describes the Param 8600 system architecture. Param is a multi-user scalable Multiple Instruction Multiple Data (MIMD) parallel computer capable of exceeding 1 Gflops of peak performance, with primary aggregate memory of 1 GByte and auxiliary storage of 20 GByte on a system with sixteen node processors [Eknath, 91]. Intel's i860 provides a peak computing power of 80 MFLOPS, but sustained computing power is less than 5MFLOPS [Murthy, 91].

4.4.1.1 Param 8600 Hardware Architecture: Each node of the Param 8600 comprises one i860 processor and four transputers. Sixteen such nodes are interconnected to form one cluster of the Param system as shown in figure 4.13. Param nodes in a cluster are connected through 96 x 96 cross point switches. Four such Param clusters can be connected through 64 x 64 way cross point switches. Therefore, a fully configured Param system can have as many as 64 nodes.

Figure 4.14 shows the architecture of a Param system node. In Param 8600, each node is equipped with four transputers T805, one i860, and a local memory. Serial communication links between two processors are called CSP (Communication Sequential Processor) channels [Ram, 91]. The communication speed of each CSP channel is 60 MBytes/sec. An important feature of this architecture is its very low interprocessor communication overhead. Another communication path between the four transputers and the i860 co-processors is the 32-bit high speed bus shown in figure 4.14. A shared bus can reduce the effective data transfer bandwidth and eventually the system performance. To avoid this bus bottleneck, C-DAC has developed an alternative scheme for data transfer. The Intel i860 co-processor and the transputers have independent memories and data is exchanged between these through the CSP channels. The i860 is treated as a computational resource for the four transputers. These four transputers can also participate in computation, but are used primarily as communication engines.

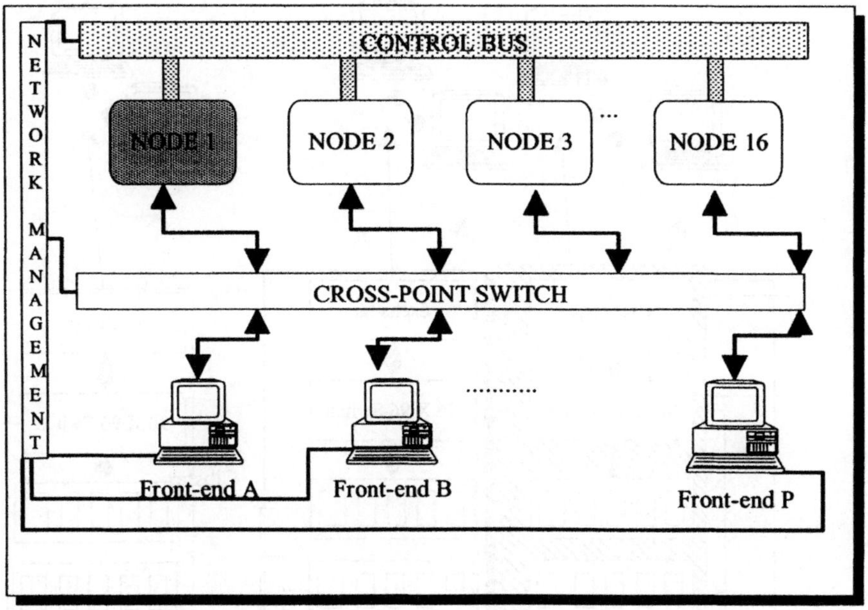

Figure 4.13a Param cluster architecture

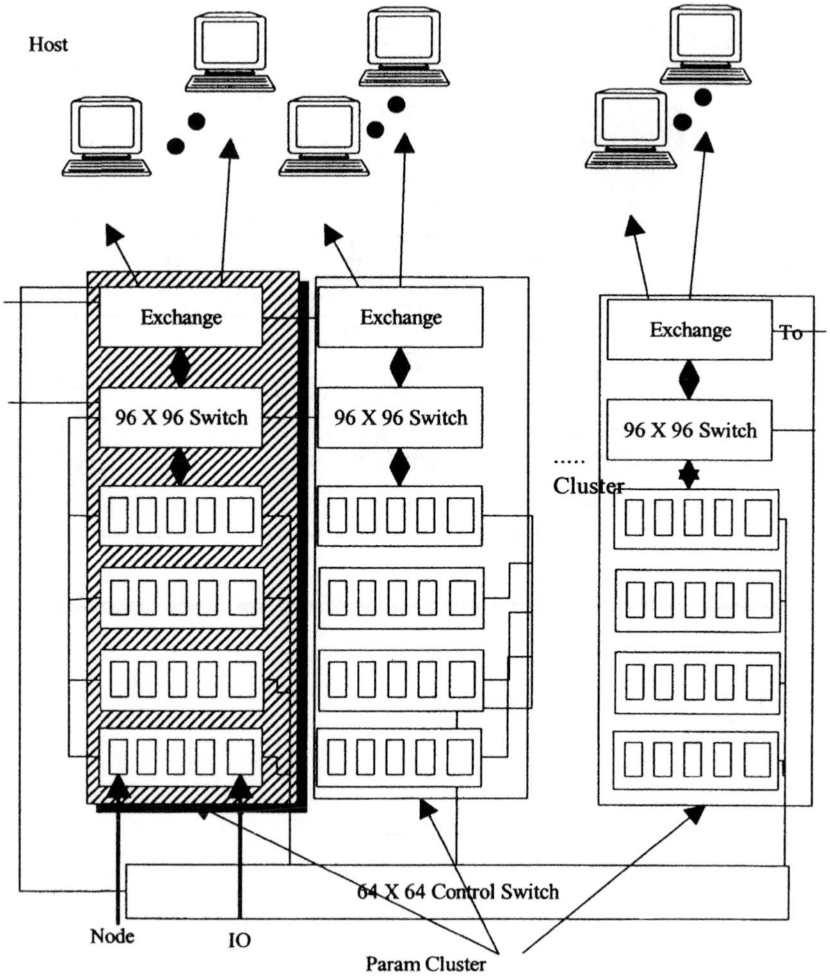

Figure 4.13b Fully configured Param system

Figure 4.13 Param system architecture

Each processor node is configured with an i860 processor, an i960CA superscalar processor, 4 - 16 MBytes of main memory, an I/O node and a Multibus II interface as shown in figure 4.15.

Apart from the local storage memory, Param is equipped with a Mass Storage System (MSS). A Disk cluster is the basic mass storage unit in the Param system. The MSS [Eknath, 91] provides high I/O bandwidth and large secondary storage capacity. The MSS consists of an array of disk drives and a number of I/O nodes with SCSI interfaces to these drives. Up to four high capacity disks are attached to each I/O node. In addition, one tape cartridge unit is provided per I/O node.

4.4.1.2 The Paras Parallel Programming Environment: Paras is a comprehensive parallel programming environment that was developed for the Param system, and similar class of message passing parallel computers. It provides a range of services to enable application software developers to utilise the hardware in a straight forward and easy to use manner. Its main components are:

- Development tools, such as compilers (for C and Fortran), linkers, debugger, librarian. It provides a rich and powerful runtime environment to the executing programs [Geetha, 91].
- Library environments such as, PARUL (PARallel User Library), for distributed memory multiprocessor systems [Kumar, 92], imagePRO for image processing [Udpikar, 91].
- Application kernels such as MTK/860 [Rao, 91]. These kernels support page-level protection, and provide call interface to user applications for performing various operations such as create, terminate, suspend, and resume.
- Interprocessor communication software PRESHAK relieves the application developers from designing and developing the communication layer for their parallel algorithms [Srivastava, 91].
- Routines for creating and managing a distributed file system.

Paras includes a configuration language which is written in the C programming language. Configuration files are named with the extension .cfs. A configuration file has three parts: hardware specification, task specification, placement of tasks on the declared processors and connectivity of the processors [Rashinker, 91]. Hardware is defined in terms of processor name and its memory. A task is specified in terms of stacksize and heapsize required for the application to run on the processor. Connectivity is specified through the links on a processor. Some of the most useful routines in the Paras programming environment are:

index_port_locate():	locate the port address to send the data.
index_port_create():	create port address to receive the data.
sync_send():	transfer the data synchronously.
block_receive():	receive the data in chunks of blocks.
get_nodeid():	get the node processor identification number.
sem_wait():	wait for the semaphore status.

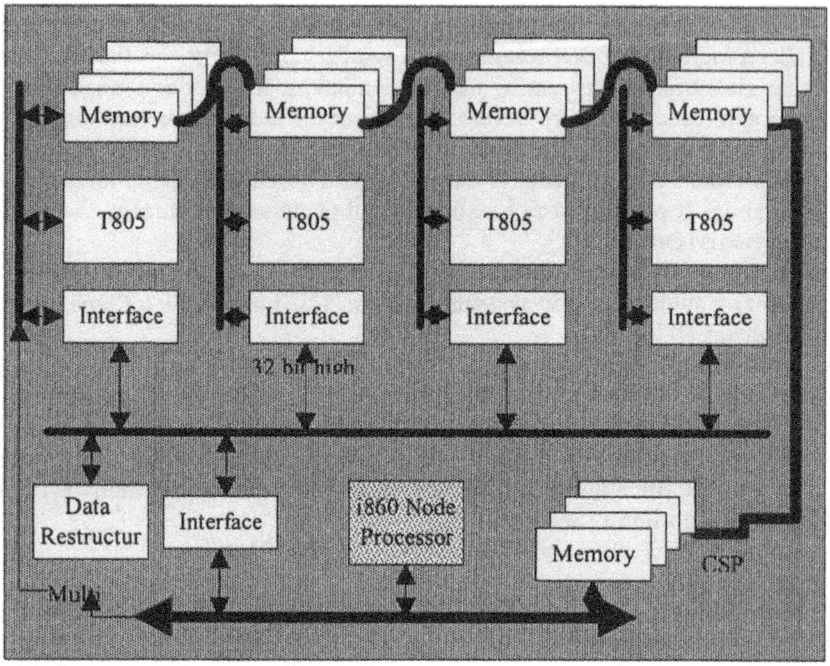

Figure 4.14 Architecture of a Param 8600 node

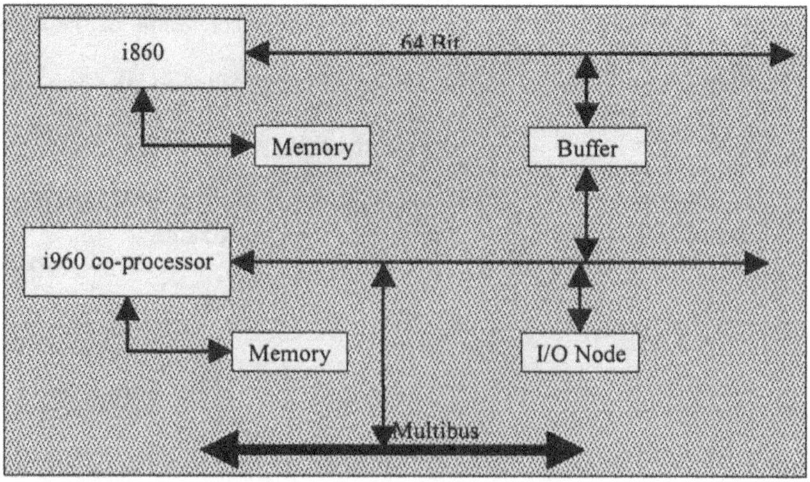

Figure 4.15 i860 node architecture

4.4.2 Implementation of the JPEG Algorithm on the Param System

The JPEG algorithm was implemented on a three node Param system [Bevinakoppa, 95]. Plans used on this architecture was represented by,

Param Plans := P(NBCT, NIPC, BBIP, DMA, DToT, NP)

where,
NP = 1 | 2 | 3 ,
Image Size = 125 x 125 | 228 x 231 | 625 x 423.

This section explains the implementation procedure and the program structure employed in implementing the JPEG algorithm on the Param system.

A Personal Computer (PC) with a transputer was used as the front-end to the Param system. A single transputer plugged into the front-end PC served as the host processor for the processing nodes on the i860 based back-end Param system. Three nodes on a single cluster were connected in a tree topology, as shown in figure 4.16. The implementation procedure used on the three node Param system is shown in figure 4.17 as a flow diagram. In this figure, the transputer on the PC is taken as the host processor and N1, N2 processors as the processing nodes.

The source image is initially stored on the host processor. Node processors wait for components of the source image to be down loaded. The host processor partitions the image into two parts and sends each part along with the required header information to node processors N1 and N2. For image partitioning a block of 8x8 samples is used as an atomic component. One third of the image is retained on the host node processor.

The distribution of image parts is shown in figure 4.18 in the form of a path graph. Image compression is performed on each processor in parallel. The encoded image is composed using a reverse procedure with respect to the procedure used for image distribution.

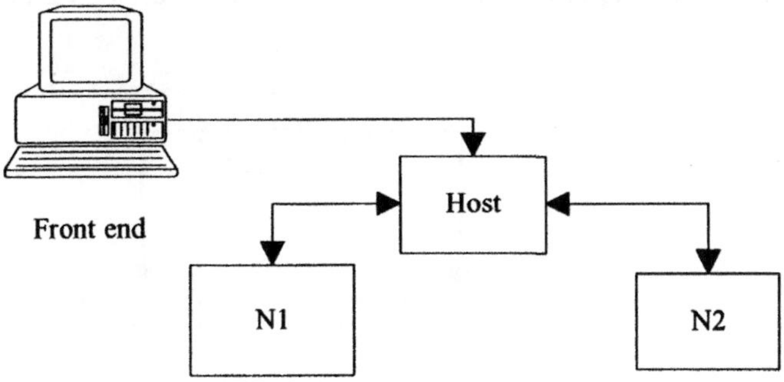

Figure 4.16 Three nodes connection in Tree topology

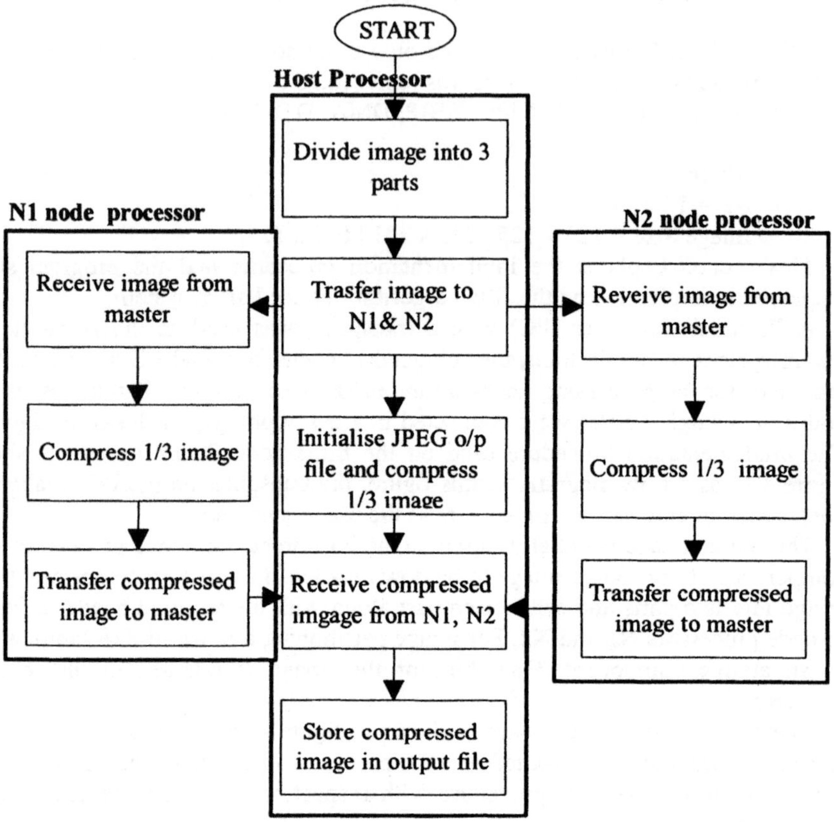

Figure 4.17 Flow diagram of implementation procedure on the Param system

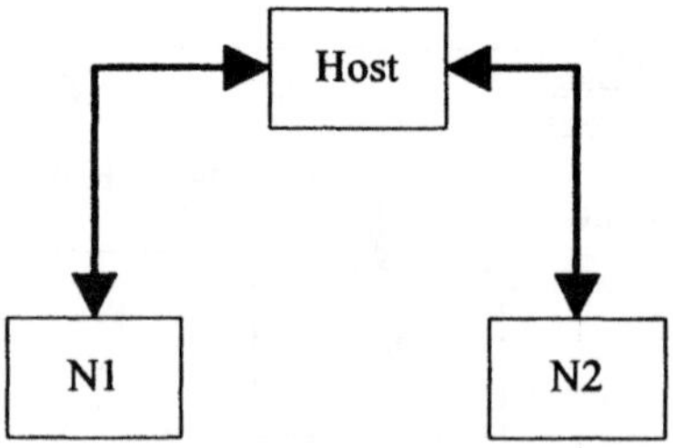

Figure 4.18 Path graph for distribution and composition of image parts

4.4.3 Experimental Results

Table 4.5 tabulates the time taken for compressing eight-bit resolution images of different sizes on the Param system for different number of node processors. Execution times on a system with one to three nodes were obtained by actual implementation, whereas execution times on four to sixteen processors were calculated with the help of Gantt charts by using execution and communication times obtained on the system with three processors.

The minimum execution times obtained for 125x125 and 228x231 image sizes are 0.038 seconds and 0.073 seconds respectively, on a system with nine node processors. Whereas, for a 625x423 image size the minimum execution time obtained is 0.327 seconds on a thirteen node Param system.

Table 4.5 Execution times of the JPEG algorithm on the Param system

Number of Processors - NP	Execution time in seconds		
	For 125x125 image	For 228x231 image	For 625x423 image
1	0.309	0.6	2.93
2	0.158	0.308	1.503
3	0.108	0.208	1.01
5	0.065	0.124	0.602
7	0.048	0.09	0.43
9	**0.038**	**0.073**	0.34
13	0.044	0.079	**0.327**
17	0.05	0.083	0.369

4.5 Performance Comparison of Parallel Computers

The performance of a parallel algorithm can be analysed in terms of speedup, scalability, and efficiency as explained in chapter 2. Section 4.5.1 gives the speedup and efficiency of the JPEG algorithm on the Mercury system. Section 4.5.2 gives the speedup and efficiency obtained on the Shiva system, and section 4.5.3 gives the speedup and efficiency for the Param system.

4.5.1 Speedup and Efficiency of the JPEG Algorithm on the Mercury System

Speedup and efficiency obtained on the Mercury system using POSIC communication routines and MPP communication routines is given in tables 4.6 and 4.7 respectively. Graphs of speedup and efficiency of the JPEG algorithm on the Mercury system are shown in figure 4.19 and 4.20. Speedup and Efficiency were defined in section 3.4 as,

$$S = T_1 / T_N,$$

$$\eta = S / N,$$

where,
S = Speedup,
T_1 = Time taken on single processor,
T_N = Time taken on N number of processor,
η = Efficiency,
N = Number of Processors.

From speedup graphs for the POSIC routines (figure 4.19a) and MPP routines (figure 4.20a), we can see that the scalability for all image sizes is close to 9. The efficiency graph for POSIC (figure 4.19b) and MPP routines (figure 4.20b) shows that efficiency of 90% or higher was obtained for up to five processors. This implies that even though there is increasing speedup values for upto nine processors, the marginal cost of adding more than five processors is rather high, In other words, from cost-benefit analysis point of view a five processor system would be most cost effective.

Table 4.6a Speedup on the Mercury system using POSIC communication routines

Number of Processors - NP	Speedup - S		
	For 125x125 image	For 228x231 image	For 625x423 image
1	1	1	1
2	1.92	1.93	1.95
3	2.84	2.89	2.92
5	4.48	4.52	4.59
7	**4.89**	**4.91**	**5.25**
9	3.65	4.02	4.31
13	1.92	2.1	2.89
17	0.05	0.083	0.369

Table 4.6b Efficiency on the Mercury system using POSIC communication routines

Number of Processors - NP	Efficiency - η		
	For 125x125 image	For 228x231 image	For 625x423 image
1	1.00	1.00	1.00
2	0.96	0.97	0.98
3	0.95	0.96	0.97
5	**0.90**	**0.90**	**0.92**
9	0.54	0.55	0.58
13	0.28	0.31	0.33
17	0.11	0.12	0.17
17	0.05	0.083	0.369

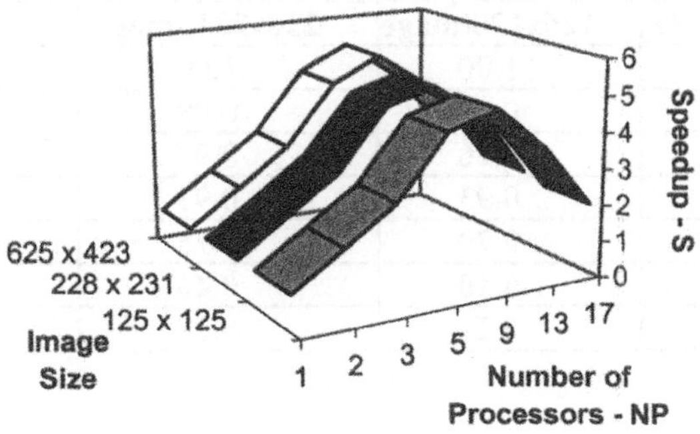

Figure 4.19a Graph of speedup on the Mercury system using POSIC communication routines

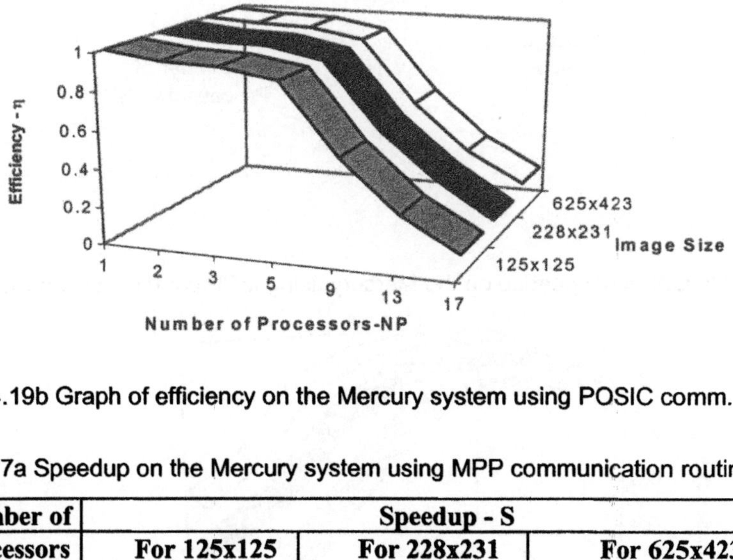

Figure 4.19b Graph of efficiency on the Mercury system using POSIC comm. routines

Table 4.7a Speedup on the Mercury system using MPP communication routines

Number of Processors - NP	**Speedup - S** For 125x125 image	For 228x231 image	For 625x423 image
1	1	1	1
2	1.93	1.95	1.97
3	2.88	2.91	2.96
5	**4.63**	**4.65**	**4.71**
9	6.3	6.77	7.29
13	5.21	5.95	6.31
17	3.57	4.76	5.44

Table 4.7b Efficiency on the Mercury system using MPP communication routines

Number of	Efficiency - η		
ProcessorsNP	125x125image	228x231 image	625x423 image
1	1.00	1.00	1.00
2	0.97	0.98	0.99
3	0.96	0.97	0.99
5	0.93	0.93	0.94
9	0.70	0.75	0.81
13	**0.40**	**0.46**	**0.49**
17	0.21	0.28	0.32

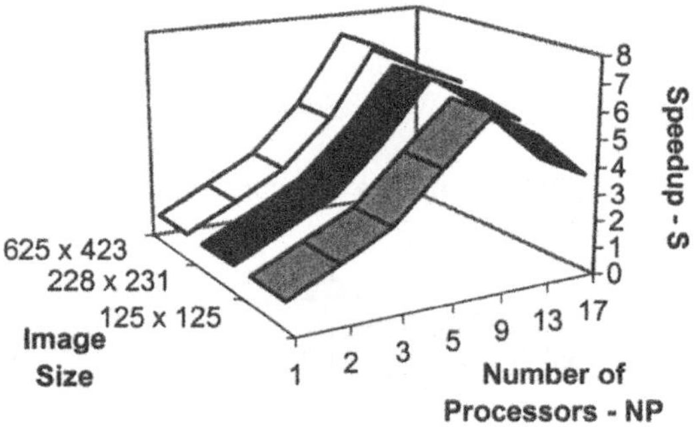

Figure 4.20a Graph of speedup on the Mercury using MPP communication routines

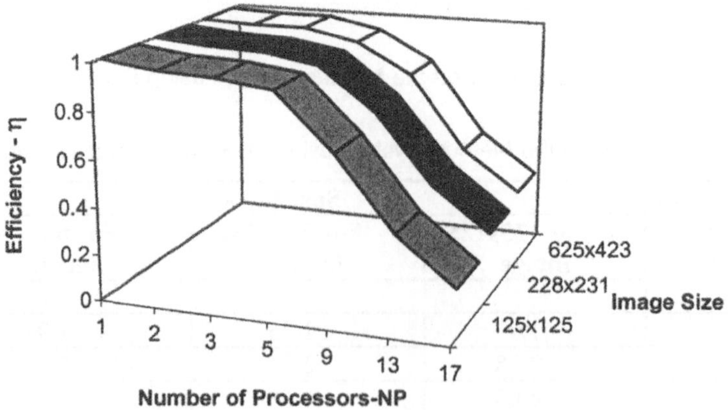

Figure 4.20b Graph of efficiency on the Mercury using MPP communication routines

Comparison of speedup and efficiency values obtained using the POSIC and the MPP communication routines for image size 625x423 are given in table 4.8 and the graphs of the same are given in figure 4.21. From the speedup graph we can see that speedup obtained using the MPP communication routines is higher than the speedup obtained by using the POSIC communication routines. Though the scaleup obtained in both cases is close to 9. The efficiency graph (figure 21b) we can see that higher efficiency can be obtained using the MPP communication routines.

Table 4.8a The speedup comparison between POSIC and MPP communication routines

Number of Processors - NP	**Speedup - S**	
	For POSIC communication routines	**For MPP communication routines**
1	1	1
2	1.95	1.97
3	2.92	2.96
5	4.59	4.71
9	5.25	7.29
13	4.31	6.31
17	2.89	5.44

Table 4.8b Efficiency comparison between POSIC and MPP communication routines

Number of Processors - NP	**Efficiency - η**	
	For POSIC communication routines	**For MPP communication routines**
1	1.00	1.00
2	0.98	0.99
3	0.97	0.99
5	**0.92**	**0.94**
9	0.58	0.81
13	0.33	0.49
17	0.17	0.32

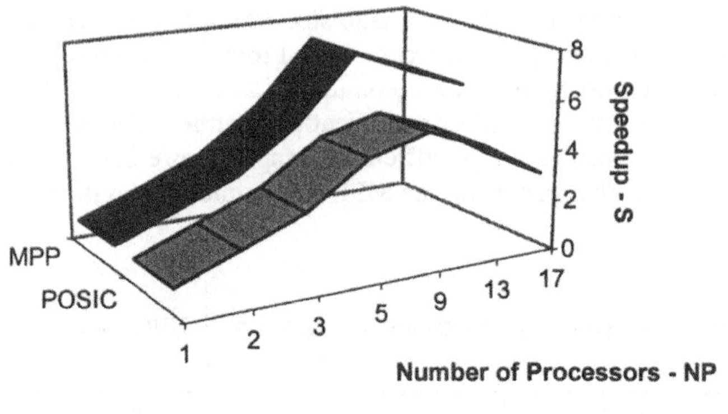

Figure 4.21a A comparison of speedup obtained on the Mercury system using the POSIC and the MPP communication routines

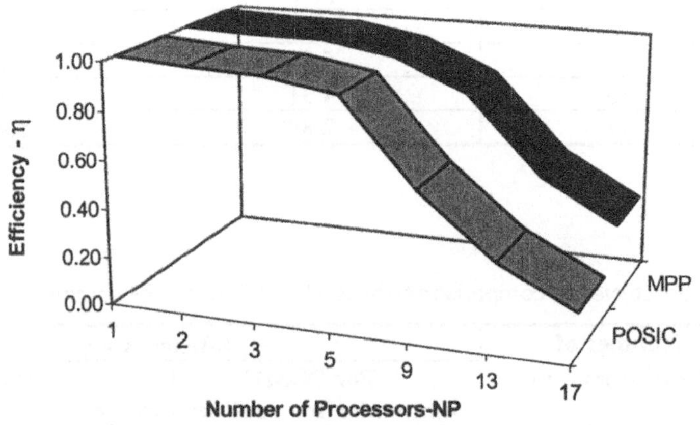

Figure 4.21b A comparison of efficiency obtained on the Mercury system using the POSIC and the MPP communication routines

4.5.2 Speedup and Efficiency of the JPEG Algorithm on the Shiva System

Speedup and efficiency values obtained on the Shiva system are given in tables 4.9a and 4.9b respectively and the graphs of the same are shown in figure 4.22a and 4.22b respectively. From the speedup graph we can see that the speedup increases as the image size increases for number of processors ≤ 3. Whereas, the speedup decreases as the image size increases for number of processors ≥ 3. Because in a shared memory architecture memory contention increases with an increase in the number of processors. Scaleup for all image sizes is 9. From the efficiency graph we can see that efficiency is higher than 50% for upto nine processors. Whereas, for upto five processors it is higher than 80%. Once again we can conclude that a system with five processors would be most cost effective.

Table 4.9a Speedup of the JPEG algorithm on the Shiva system

Number of Processors - NP	Speedup - S		
	For 125x125 image	For 228x231 image	For 625x423 image
1	1	1	1
2	1.86	1.96	1.98
3	2.54	2.69	2.91
5	3.89	3.98	4.61
7	4.67	4.82	5.06
9	**4.8**	4.96	**5.31**
13	4.79	4.35	4.4
17	3.06	2.89	2.72

Table 4.9b Efficiency of the JPEG algorithm on the Shiva system

Number of Processors - NP	Efficiency - η		
	For 125x125 image	For 228x231 image	For 625x423 image
1	1.00	1.00	1.00
2	0.93	0.98	0.99
3	0.85	0.90	0.97
5	**0.80**	**0.88**	**0.92**
7	0.67	0.69	0.72
9	0.53	0.55	0.59
13	0.37	0.33	0.34
17	0.18	0.17	0.16

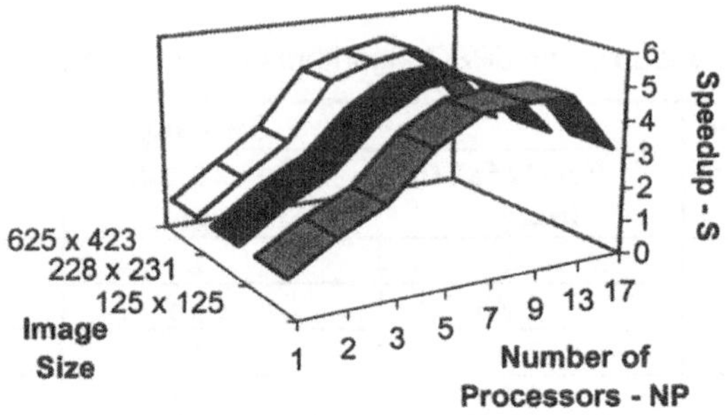

Figure 4.22a Graph of speedup on the Shiva System

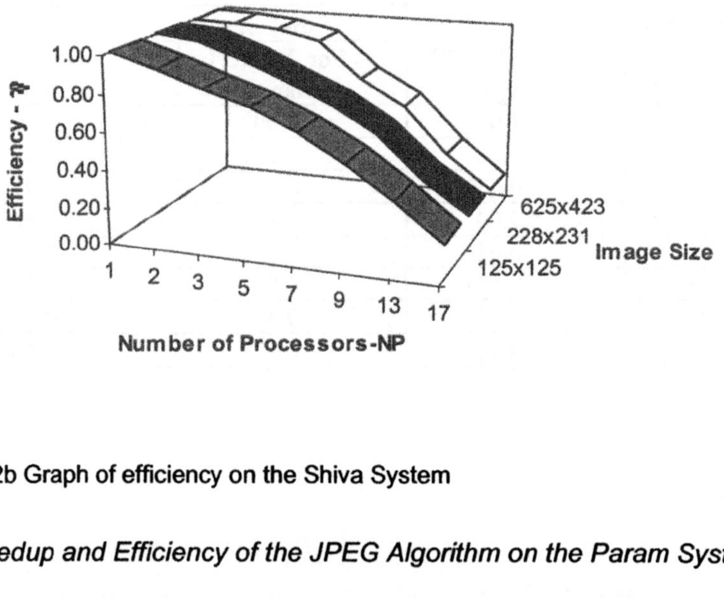

Figure 4.22b Graph of efficiency on the Shiva System

4.5.3 Speedup and Efficiency of the JPEG Algorithm on the Param System

The speedup and efficiency values obtained on the Param system are given in tables 4.10a and 4.10b respectively and the graphs of the same are shown in figures 4.23a and 4.23b respectively. Speedup depends mainly on computation time with respect to processors and communication time between processors. Communication time can be reduced by transferring more number of bits in a single frame, thereby reducing the communication overhead. Therefore, as we can see in figure 4.23a, speedup increases with increase in image size. Scalability for 125 x 125 and 228 x 243 image size is nearly 9, where as for a 625 x 423 image it is nearly 13. From the efficiency graph (figure 4.23b), we can see that the efficiency is higher than 50% for upto thirteen processors for 125 x 125 and 228 x 243 image sizes, and for 625 x 423 image size it is seventeen processors. Whereas, for upto nine processors it is higher than 80%. From this we can conclude that a system with nine processors would be cost effective.

Table 4.10a Speedup of the JPEG algorithm on the Param system

Number of Processors - NP	Speedup - S		
	For 125x125 image	For 228x231 image	For 625x423 image
1	1	1	1
2	1.95	1.95	1.95
3	2.86	2.88	2.90
5	4.75	4.82	4.87
7	6.5	6.64	6.81
9	**7.93**	**8.17**	8.63
13	7.1	7.58	**8.97**
17	6.34	7.21	7.93

Table 4.10b Efficiency of the JPEG algorithm on the Param system

Number of Processors - NP	Efficiency - η		
	For 125x125 image	For 228x231 image	For 625x423 image
1	1.00	1.00	1.00
2	0.98	0.98	0.98
3	0.95	0.96	0.97
5	0.95	0.96	0.97
7	0.93	0.95	0.97
9	**0.88**	**0.91**	**0.96**
13	0.55	0.58	0.69
17	0.37	0.42	0.47

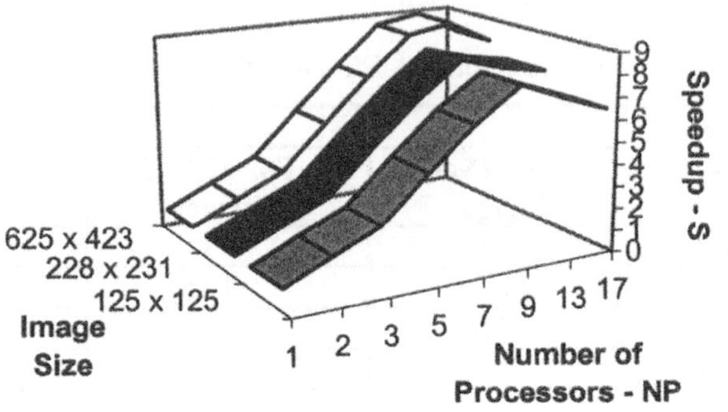

Figure 4.23a Graph of speedup on the Param system

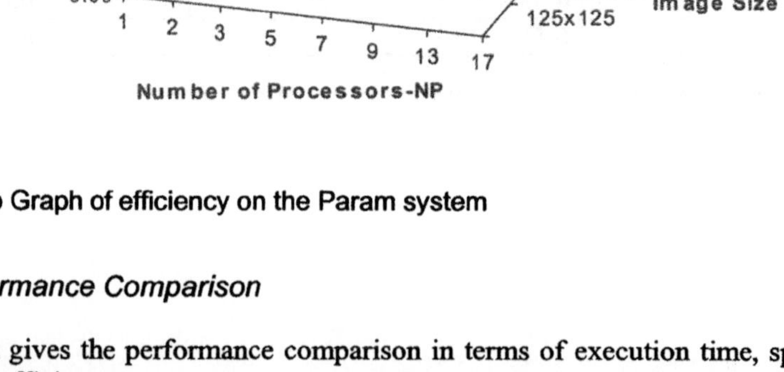

Figure 4.23b Graph of efficiency on the Param system

4.5.4 Performance Comparison

This section gives the performance comparison in terms of execution time, speedup, scaleup and efficiency.

Execution times obtained on the three parallel computers for the 625 x 423 image size are given in table 4.11. From this table we can see that the execution time on the Mercury system are much higher as compared to those obtained on the Shiva and the Param systems. The execution times obtained on the Param system are the lowest.

Table 4.11 Execution times of the JPEG algorithm on the three parallel computers

Number of Processors - NP	Execution times in Seconds		
	For Mercury System	For Shiva System	For Param System
1	55.86	7.02	2.93
2	28.65	3.54	1.503
3	19.13	2.41	1.01
5	12.17	2.98	0.60
9	**10.64**	**2.26**	**0.34**
13	12.96	2.96	0.327
17	19.33	2.58	0.369

The speedup and efficiency values obtained on the three parallel computers for the 625 x 423 image size is tabulated in tables 4.12a and 4.12b respectively and the graphs of the same are shown in figures 4.24a and 4.24b respectively. From the speedup graph we can see that the speedup values obtained on all three parallel computers are very similar for NP ≤ 3. The speedup for NP > 3 on the Mercury system is lower as compared to the other two architectures. For NP $= 17$ the speedup is lowest on the Shiva system. From this we can conclude that the addition of processors lead to increased bus contention and reduction in the speedup obtained on the Shiva system. The speedup for the Param system is higher as compared to other architectures. Scaleup for Mercury system and for Shiva system is 9, and for the Param system it is equal to 13.

From the efficiency graph (figure 4.24b) we can see that the efficiency is higher than 50% for upto nine processors on Mercury and Shiva systems and on Param system the efficiency is higher than 50% for upto thirteen processors. Whereas, for upto five processors it is higher than 90% on Mercury and Shiva systems and for upto nine processors the efficiency is higher than 90% on Param system. From this we can concluded that Param system with nine processors would be most cost effective.

From speedup, efficiency and scaleup figures discussed above we can conclude that the JPEG algorithm performs better on a hybrid memory architecture.

Table 4.12a Speedup of the JPEG algorithm on the three parallel computers

Number of Processors - NP	Speedup - S		
	For Mercury System	For Shiva System	For Param System
1	1	1	1
2	1.95	1.98	1.95
3	2.92	2.91	2.90
5	4.59	4.61	4.87
9	**5.25**	**5.31**	8.62
13	4.31	4.4	**8.96**
17	2.89	2.72	7.94

Table 4.12b Efficiency of the JPEG algorithm on the three parallel computers

Number of Processors - NP	Efficiency - η		
	For Mercury System	For Shiva System	For Param System
1	1.00	1.00	1.00
2	0.98	0.99	0.98
3	0.97	0.97	0.97
5	0.92	0.92	0.97
9	**0.58**	**0.59**	0.96
13	0.33	0.34	0.69
17	0.17	0.16	**0.47**

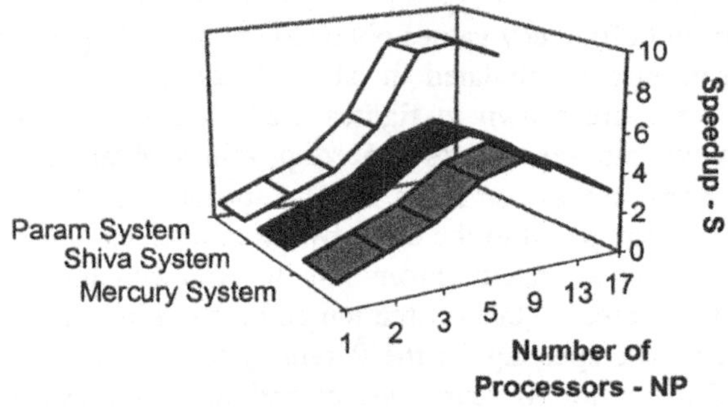

Figure 4.24a Speedup graph for three parallel computers

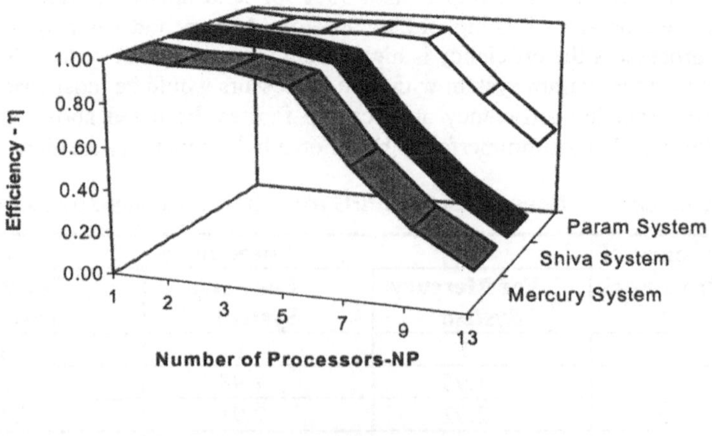

Figure 4.24b Efficiency graph for three parallel computers

4.6 Summary

This chapter described the hardware architecture and implementation of the JPEG algorithm on three parallel architectures. Experimental results, speedup and efficiency graphs were given in this chapter. Speedup depends mainly on computation time on the processors and the communication time between the processors. The communication time can be reduced by transferring more bits in a single frame, thereby reducing the communication overhead. Therefore, in all speedup graphs, speedup increases with increase in image size.

On the Mercury system, a comparison of POSIC and MPP based communication shows that the speedup and efficiency obtained with the MPP communication routines is higher than that obtained with the POSIC communication routines. But, the POSIC communication routines provide greater functionalities than the MPP communication routines.

A comparison of the three architectures showed that the hybrid memory architecture, such as the Param system, gives the best performance in terms of speedup, scaleup and efficiency.

5 SIMULATION OF DIGITAL IMAGE COMPRESSION TECHNIQUES

5.1 Introduction

This chapter describes modelling and simulation methods used for simulating parallel processing of image compression techniques using the Network II.5 simulation package.

Model building and simulation involves ten steps, viz. problem formulation, model building, data collection, model translation, model verification, model validation, experiment planning, experimentation, analysis of results, and documentation. Each of these steps were described briefly in the third chapter. Image compression Plans on different parallel computer architectures have been modelled using the Network II.5 simulation package. Details of the model building process and the process of running simulation experiments for various Plans are given. Simulation results are compiled to evaluate the performance of these Plans.

Speedup, scaleup and efficiency of each Plan is given, and the performance of different Plans is compared in terms of speedup, efficiency and scaleup.

Predicting the performance of digital image compression algorithms on different parallel architectures is a complex problem; simulation techniques can therefore assist in the evaluation process.

To systematise the modelling and simulation work the idea of a Plan is introduced, in section 3.3. Each Plan was translated into a model using the Network II.5 package. The procedure adopted for developing the models, and running the simulations is explained in section 5.2.

Performance figures for the various Plans were obtained by running simulations on their respective models. Performance figures are derived from the execution times obtained from the simulation experiments. Execution times of the various models are given in section 5.3. Performance of these models, in terms of speedup, scaleup and efficiency is discussed in section 5.4.

5.2 Simulation Procedure

This section describes the model building and simulation procedure for image compression techniques on parallel computers using the Network II.5 simulation package. Procedure for model building and simulation includes problem statement, model building, system simulation, system analysis and validation. These steps are described in the following sections.

5.2.1 Problem Statement

The first step in building a model is to make a clear statement of the problem. In this research the main issue is to study the various options for implementing digital image compression techniques on parallel computers. This involves the study of digital image compression techniques as well as parallel computer architectures. There are various options for implementing digital image compression on parallel computers. Each possible implementation is described as a Plan in this book.

Model building and simulation on the Network II.5 simulation package is described in this section by taking the Plan given in equation 5.1 as an example. In this Plan three Processing Elements are connected in a tree topology. Thus taking,

P1 = P(BCT, NIPC, BWIP, DMA, DTrT, 3) (5.1)

where,
P is the Plan for implementation,
BCT is the Block Comparator Technique used for image compression,
NIPC is Non-Inter-Processor Communication method used in conjuction with Block Dependency,
BWIP is Balanced Workload Image Partitioning method,
DMA is Distributed Memory Architecture,
DTrT is Distributed memory architecture using Tree Topology,
NP - Number of Processors = 3.

5.2.2 Model Building

The steps involved in model building are: create network topology, define system operations and model verification. These steps are described in the following sections.

5.2.2.1 Create Network Topology: The first step in model building is to create the network topology. In Network II.5 the hardware model can be represented graphically as a collection of devices such as Processing Elements (PE), Transfer Devices (TD), and Storage Devices (SD) as explained in chapter 3.

For Plan P1 given in equation 5.1, three Processing Elements (PE), one Storage Device (SD) and three Transfer Devices (TD) are required as shown in figure 5.1. In this figure Storage Device SD-1 is connected to the Host Processor through the TD-1 Transfer Device. The Host Processor is connected to PE-1 through TD-2, and to PE-2 through TD-3.

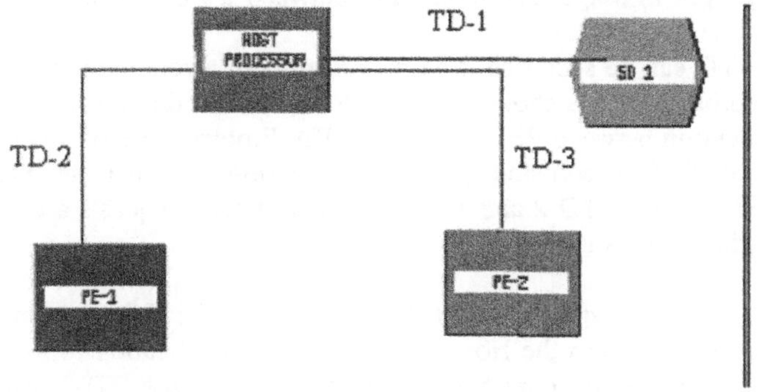

Figure 5.1 Graphical representation of Plan P_1

5.2.2.2 Define System Operation : The steps involved in defining the system operations are:

1. Specify the characteristics of each hardware devices.
2. Define instructions for each PE.
3. Construct the software modules.

These three steps are described in the following sections.

<u>**Specify the characteristics of hardware devices:**</u> Each hardware component is specified by a set of parameters. The parameters selected for Processing Elements, Transfer Devices, and Storage Devices are as follows. In this section Intel's i860 processor was used as the CPU being modelled, because the i860 based Shiva machine was used as one of the parallel processors for implementing the JPEG algorithm.

- **PE specification**: A PE is characterised by its Basic Cycle Time, Instruction Repertoire, Message List Size, and Overheads. Three PEs are named as Host Processor, PE-1 and PE-2. Screen outputs showing the details of the Host Processor and PE-1 are shown in figure 5.5. The Basic Cycle Time for these three PEs were specified as one micro second. The computation time for instructions defined for each PE are taken from the experimental results. Basic Cycle Time is the basic time unit on which the execution time of all processing instructions of a PE are built .

Host Processor, PE-1, PE-2:

Basic Cycle Time = 1 MIC[1] (Micro Seconds)

- **TD specification**: TDs are the links connecting PEs and SDs to move the data either between two PEs or between a PE and an SD. Data is moved between two

[1] MIC is the abbreviation used for microseconds in the NETWORK II.5 system.

PEs by a message instruction and between a PE and an SD, as the result of a read/write instruction.

TD can be specified in terms of protocol, cycle time, bits/cycle, cycle/word, word/block etc. as shown in figure 5.6. Protocol defines the method of resolving contention between PEs for a single TD. Protocols are of types; First Come First Serve (FCFS), collision, priority, token ring, crossbar etc. FCFS protocol was used for TD-1, TD-2 and TD-3. In this protocol, requests are served in the order in which they arrive.

Two TD types were used in this Plan. The first type of TD was used to connect the SD to the Host Processor. The specification form for TD of the first type is shown in figure 5.6a. TD of the second type, TD-2, was used to connect the two PEs. TD-2 specification form is shown in figure 5.6b. Specified parameters for these TD are as follows.

TD-1: TD between SD and Host Processor:

Protocol	=	FCFS
Cycle time	=	0.5 MIC
Bits / Cycle	=	64
Cycle / word	=	1
Word / block	=	1

TD-2 and TD-3: TDs between two PEs:

Protocol	=	FCFS
Cycle Time	=	1 MIC
Bits / cycle	=	64
Cycle / word	=	1
Word / block	=	1
Block overhead	=	0.05 MIC

- **SD Specification**: Storage devices contain user named files and unstructured storage. In our case, SD-1 contains two files named as Source Image File and Compressed Image File. Read and write instructions are used to access these files. When a read instruction references an SD, it checks for the Source Image File to see if the requested file is available. If it is available, the file is read into the PE.

  When write instruction attempts to put a file into an SD, it checks to see if there is enough space available. If required space is available it accepts the file. If adequate space is not available, a warning message is issued to the user.

  Files are read or written analogous to the way real storage devices work. Storage devices automatically decompose all file reads and writes into words and blocks. Every SD can be specified by attributes such as Name, Read access time, Write access time, Read word overhead time, Write overhead time, bits/word, number of ports, word / block, file list, as shown in figure 5.7. Following are the some of the attributes used for SD-1.

SD-1:

Capacity: Total number of bits that SD can hold = 1 GBits
Bits/word = 64
Number of Ports = 1
Word / block = 1
File List = 2 (Source Image File, Compressed Image File)

Define instructions for each PE: The next step of model building is to define the instructions for each PE. The Instruction Repertoire consists of four types of instructions. That is, processing instructions, read/write instructions, message instructions and semaphore instructions. Instructions for a Plan are divided as instructions on the Host Processor, and instructions on PE-1 and PE-2. These instructions are described in the following sections.

For Plan P1 image size was taken as the 1100 x 900 samples, Number of Similar Blocks (NSB) = 10%. From this data the parameters are calculated as follows.

Image Size = 1100 x 900 samples
= 1100 x 900 bits (For monochrome image @ 1 bit / sample)

$$\text{Number of Blocks (NB)} = \frac{1100}{8} * \frac{900}{8} = 15594$$

Number of Similar Blocks (NSB) = 10% of NB = 10 % of 15594 ≅ 1560

Instructions on the host processor are shown in figure 5.5a and are explained below.

Host Processor Instructions:

- **Start Time:** This is a semaphore instruction to set the '**Start Time**' semaphore. This semaphore is used to measure the execution time for each experiment. It is set at the beginning of the experiment and reset at the end. Thus the execution time is simply the length of time for which this semaphore stays set.
- **Receive Img from SD:** This is a read instruction. This instruction reads the source image file from the SD. Time taken to receive the image file from SD depends on the number of bits to be received. The number of bits received from SD for Plan P1 = 1100 x 900 = 990000 bits.
- **Block Comparison:** This is a processing instruction where the number of cycles were specified to process block comparison operations. Time taken for block comparison was taken from experimental results and it depends upon the image size. For the 125 x 125 image size time taken for Block Comparison = 7300 MIC, for a 625 x 423 image size it is = 28000 MIC, for a 1100 x 900 image size it is = 390000 MIC. For Plan P1 the number of cycles = 390000 for the 1100 x 900 image size.
- **Partition Image:** This is a process instruction. Partitioning time is quite small, about 50 MIC. The partition step partitions the image blocks into three parts. Here the Balanced Workload Image Partitioning (BWIP) method was used. In this step, for Plan P1, the Number of Unique Blocks was taken as 15594 - 1560 =

14034. Therefore, all three processors get 14034 / 3 = 4678 blocks for processing, assuming that all blocks are of equal intensity value.

- **Send Imgpart-1 to PE-1:** This is a message instruction. The time taken to transmit part of the image from the host processor to PE-1 depends upon the number of bits in image part-1. In this example NIPC method was used for Block Dependency method. Therefore the number of bits to be transmitted to PEs includes part of the image and the neighbouring samples. Part of the image consists of number of blocks x number of bits in one image block. Therefore, the number of bits transmitted = Number of Blocks x bits/ image block + neighbouring samples = 4678 x 8 + 2200 = 37424 + 2200 = 39624 bits.
- **Send Imgpart-2 to PE-2:** Details of this instruction are similar to those of the **Send Imgpart-1 to PE-1** instruction.
- **Process Image Block:** This is a processing instruction. It represents the compression operation carried out on each image block. The time taken for processing a complete image ($T_{procimg}$) is equal to number of blocks x time taken for processing one block as given in the following equation as,

$$T_{procimg} = \text{Number of Blocks (NB)} \times T_{procblock} \quad (5.2)$$

where, $T_{procblock}$ is the time taken for processing one image block.

The number of cycles for processing one image block is defined as a normal statistical distribution function (SDF). The upper limit for this normal distribution = 425 MIC, the lower limit = 475 MIC, the Standard Deviation = 18.074, and the Mean = 450 MIC. The parameters for this SDF were obtained from experimental. results.

- **Send CompImg to SD:** This is a Write instruction. It represents the operation in which after completion of compression process, compressed image data is sent to the SD. The Compressed Data Structure -3 was considered to calculate the number of bits transmitted as given in equation 5.5. Details of this equation were explained in section 2.4.2. The definition of which is repeated here as,

$$S_{BCT3} = S_{JHI} + NL * (S_{SBM} + S_{UBNF} + nl * S_{BNF}) + S_{UBM} + NUB * \frac{S_{Blk}}{BCF} + S_{EOI} \quad (5.3)$$

where,

S_{BCT3}	=	BCT Compressed image size for CIDS-3 in Bytes,
S_{JHI}	=	Size of the JPEG Header Information in Bytes = 173 bytes,
NL	=	Number of Similar Block Lists in SBG = 1,
nl	=	Number of Block Numbers in each Unique Block list = 1560,
S_{SBM}	=	Size of the Similar Block Marker = 1 byte,
S_{UBNF}	=	Size of the Unique Block Number Field = 2 bytes,

S_{UBM}	=	Size of the Unique Block Marker = 1 bytes,
NUB	=	Number of Unique Blocks = 14034,
S_{Blk}	=	Size of one block in Bytes = 1,
S_{EOI}	=	Size of the End Of Image marker = 1 byte,
BCF	=	Block Compression Factor = 6,
S_{BCT3}	=	173 + 1 * (1 + 2 + 1560 * 2) + 1 + 14034 * 8 / 6 + 1
	=	22011 bytes
	=	22011 * 8 = 176088 bits.

- **End Time:** This instruction resets the **'Start Time'** semaphore. Network II.5 determines the active duration of the **'Start Time'** semaphore from set and reset condition.

Figure 5.5b shows the instruction list for PE-1. Instructions on PE-2 are the same as the instructions on PE-1. These instructions are as follows.

PE-1 and PE-2 instructions:

- **Process Image Block:** This is a processing instruction. The number of cycles for this instruction is the same as for the Host Processor.
- **Send CompImg to Host:** This is a message instruction. It represents the operation in which compressed image is sent to the Host Processing Element. The number of bits specified = 4678 * 8 / 6 = 6238 bits.

<u>**Construct modules:**</u> Tasks to be performed by PEs are specified as modules. Module description consists of four parts, these are: scheduling conditions, processing element options, a list of instructions to execute and a list of modules to execute when this module completes. A module constantly checks its preconditions to see if the user defined scheduling criteria are met. Once all preconditions have been met the module takes the list of PEs on which it will run. A module begins execution by issuing instructions from its instruction list. Once all instructions have executed successfully it choses its successor module.

There are three modules defined for the example Plan P1, these are ProcessImg on Host module, ProcessImg on PE-1 / 2 module and Send CompImg to SD module. These three modules are shown in figures 5.8, 5.9 and 5.10 respectively.

From figure 5.8 we can see that the precondition for ProcessImg on Host module is start time = 0 MIC. Therefore, this module starts execution as soon as simulation begins. Then it issues instructions from the instruction list, one by one. First instruction in the list sets the **Start Time** semaphore. The second instruction carries out **Block Comparison**. Next, the image is partitioned into three parts, and the Host Processor sends Image Part-1 to PE-1 and Image Part-2 to PE-2. Then compression on image blocks takes place. The number of blocks to be compressed is defined by the iteration list. In this case number of iterations were defined for the ProcessImg Block instruction as 4678. This means that the **Process Image Block** instruction executes 4678 times. (In the module form, shown in figure 5.8 and 5.9, only the first three digits for the number of iterations can be seen.) Then this module chooses its successor. The successor module is Send CompImg to SD module.

The next module to execute is Send Complmg to SD as shown in figure 5.10. In this module the precondition is that messages Complmg Part-1 and Complmg Part-2 should have been received.

When Processlmg on Host module executes, the Processlmg on PE-1 / 2 module also executes parallelly. Processlmg on PE-1 / 2 module waits for the message Imgpart. When Processlmg on Host module sends Imgpart message, the Processlmg on PE-1 / 2 module takes PE-1 and PE-2 from the processor list to execute the instructions on these PEs. The first instruction in this module is **Process ImgBlock** as shown in figure 5.9. This instruction executes 4678 times because the number of iterations were defined as 4678. Then these two PEs send their respective Complmg Parts to the Host by executing **SendCompimg Host** instruction, with Complmg Part-1 and Complmg Part-2 as the message names.

As soon as the Send Complmg to SD module receives these two messages, instructions on the Host Processor start executing. This module has anded predecessors as Host Processor. Because Host processor has successor as Send Compimg to SD module. The first instruction in this module is **Send Complmg to SD** as shown in figure 5.10. It send the compressed image to the SD. Then it resets the **Start Time** semaphore by executing the **End Time** instruction.

5.2.2.3 Model Verification Model verification is assisted by the module diagram generated by Network II.5. In this display, modules are represented in a flow chart format as shown in figure 5.2. The name of the module is written inside the box. Preconditions are displayed on the right upper corner of the box. The instructions for this module are listed on the right side of the box. Output of the instructions or the messages are indicated on the right bottom corner of the box. The successor is indicated by a down arrow leading to the other box.

Figure 5.2 shows that the precondition for Processlmg on Host module is **Start Time** = 0 MIC. This module executes on the Host Processor, indicated below the precondition. The instruction list is given below the name of the processor. For Processlmg on Host module the instructions are **Start Time, Receive Img from SD, Partition Image, Send ImgPart-1 to PE-1, Send ImgPart-2 to PE-2, and Process ImgBlock.** The output of this module is ImgPart message and the **Start Time** Semaphore being set. These are displayed in the bottom right corner of the module box. The Anded successor module is Send Complmg to SD module.

For the Send Complmg to SD module the precondition is Complmg Part-1 and Complmg Part-2. The instructions for these modules are **Send Complmg to SD** and **End Time**. Output of this module is to reset the **Start Time** semaphore.

For the Processlmg on PE-1 / 2 module the precondition is ImgPart. The instructions are Processlmg Block and Send Complmg to Host. The output of this module is Complmg Part-1 and Complmg Part-2.

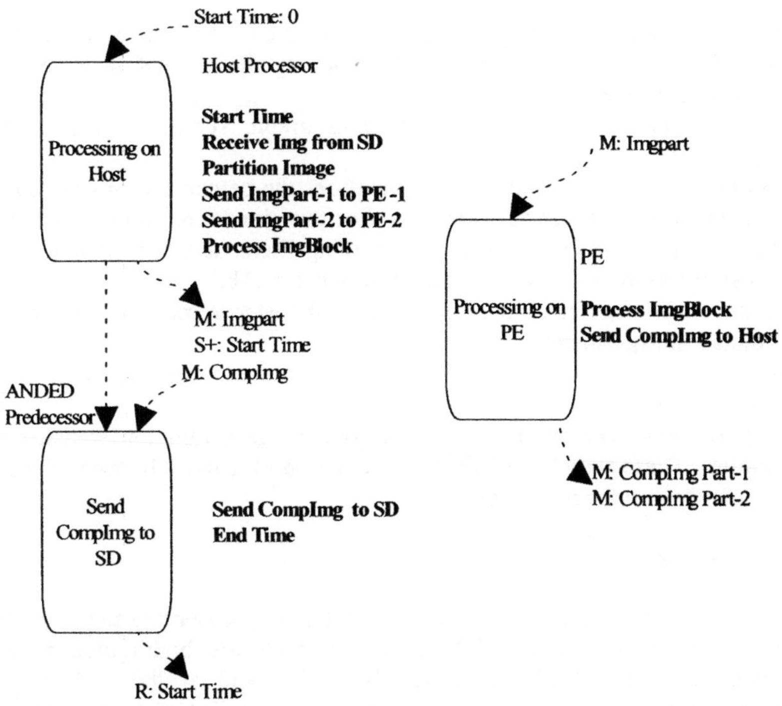

Figure 5.2 Modules diagram

5.2.3 System Simulation

After Verification, the model is subjected to a series of simulation runs. The steps involved in system simulation are, specify the run parameters, and run simulation. These are described in the following sections.

5.2.3.1 Specify Run Parameters: Simulation is run after specifying parameters such as Run length, Periodic Reports, Final reports, Trace etc. Parameters are specified in the Run Parameters form shown in figure 5.11a. These are described below:

- **Run length:** The run length represents the simulated time for which Network II.5 should run the simulation. Specified here as 6 Seconds.
- **Periodic reports:** The number of Periodic Reports were specified as 4. This gives four reports while the simulation is running. Required Periodic Reports are specified in the Periodic Reports list given at the right bottom corner of the menu. Reports for ProcessImg on Host module, ProcessImg on PE-1 / 2 module, Send CompImg to SD module and TD status were specified.
- **Final reports:** These are the reports included in the final set of reports at the end of the simulation. In these simulation the final reports for the Host processor and PE-1 were observed.

5.2.3.2 Run Simulation: A simulation experiment can be run on a model after specifying the simulation run parameters. After running the simulation experiment following facilities can be accessed.

- **Runtime reports:** This gives the utilisation graph, runtime warnings and summary reports.
- **Utilisation graph:** The utilisation graph measures the percentage of time during an interval that a PE, TD, or SD was busy. For this simulation utilisation graphs was asked for the Host Processor and Processing Element-1. Utilisation graphs for the Host Processor and PE-1 are shown in figure 5.11b.
- **Trace reports:** The runtime trace reports allow the user to monitor the progress of the simulation experiment.
- **Runtime warnings:** Runtime warnings notify potential errors as these occur during the simulation run.
- **Summary reports:** The summary reports contain the simulation statistics for model entities. These statistics encompass the period of activity between the start of the simulation and the reset time.

5.2.4 System Analysis

In Network II.5 the operation of a model can be analysed by observing the animation screens. During an animation, model hardware elements are highlighted as they become active. Icons are displayed at the beginning and at the end of data transmission, moving first from source to TD and then from TD to destination. The procedure for creating and running animation screens is as follows.

5.2.4.1 Animation: Animation consists of three steps, namely, set animation parameters, start the animation and observe the animation. These are described in the following sections.

Set animation parameters: Animation parameters can be set in the animation menu as shown in figure 5.12. This figure indicates that the animation will be done in the single step model, animation is displayed by event, and the delay time from each event is 1 sec, start time = 0 MIC.

Start animation: After specifying the animation parameters, animation is started by clicking the OK button. Network II.5 starts the animation by displaying all the elements in gray colour initially, then a red icon starts moving through, as the operation progresses.

Observe the animation: As the model's hardware elements become active, they are displayed with their characteristic colour. When a module runs on a PE, the name of the module is displayed in place of the PE name. If PE transmits a message over a TD, the PE is highlighted with its characteristic colour, the TD colour changes to the source PE colour, the message is displayed near the TD and the TD connection to the destination PE changes to the source PE's colour. Therefore, animation helps in visualising the operation of the model and locating any errors in the model's operation.

5.2.4.2 Plotting: Network II.5 simulation package provides the facility to generate utilisation and timeline status plots for PEs, TDs, SDs, modules and semaphores. Plotting consists of two steps: set the plot parameters, and plot the required data.

Set plot parameters: Prior to starting a plot the required parameters have to set as shown in figure 5.13, by entering the plot type as 'Time-Line', the list of items to plot as PEs and semaphores, and plotting Time-Span by specifying start at 0 second and end at 2.6 seconds.

Plot: Two types of plots can be produced in Network II.5: Time-Line status plots and utilisation plots. These are explained in the following sections.

1. **Time-line status plots**: The Time-Line status plot measures each device's pattern of activity during the competed simulation as a function of time. These plots can be used to compare the time spent on communication and computation. Figure 5.14 shows the Time-Line status plot for modules, PEs, TDs, and Semaphore. From this plot we can clearly see the idle time of PE-1 and PE-2 in the beginning.

2. **Utilisation plots**: The utilisation plot is a histogram which measures a device's percentage utilisation during a completed simulation as a function of time. Figure 5.15 shows the utilisation of the Host Processor, PE-1, and PE-2.

5.2.5 Validation

A model can be validated by comparing the simulation results with the implementation results. The execution times obtained from simulation experiments and those obtained by real-time implementation are given in table 5.1. Match between the real-time and simulation results is expressed as a Match Ratio defined in the following equation,

$$\text{Match Ratio} = \text{Execution Time (simulation)} \;/\; \text{Execution Time (implementation)} * 100. \tag{5.4}$$

Table 5.1 Comparison of execution times obtained from simulation and implementation for Plan P1

Number of Processors-NP	Execution Time in Sec. (Simulation)	Execution Time in Sec. (Implementation)	Match Ratio %
1	6.3828	6.7	95.26
2	3.315629	3.5	94.73
3	2.52244	2.71	93.08

From table 5.1 we can see that the values obtained from simulation are within 90% of the values obtained from experimentation. Therefore, we can say that the model for Plan P1 is a valid model. A validated model can be used for modelling Plans.

5.3 Simulation Results of Digital Image Compression Techniques

Digital image compression techniques can be implemented on parallel computers in many different ways. A specific way of implementation is called a 'Plan'. The specific Plans selected for modelling and simulation are described in section 5.3.1. Simulation results for the selected Plans are discussed in section 5.3.2. Tables giving the execution times for the various simulated Plans are given in appendix A; table A.1 to A.15. Speedup values calculated from the execution time values are also given in appendix A; tables A.21 to A.35. Speedup and efficiency graphs are plotted next to each speedup table; figures A.1 to A.15. Comparisons of the results obtained for the different Plans are given in section 5.3.2.

5.3.1 Plans Selected for Simulation

A Plan (Px) can be represented as a 6-tuple given in equation 5.5. Thus,

$$Px = P(ICT, BD, IPM, MA, MO/NT, NP) \tag{5.5}$$

where,
Px is Plan-x for implementation,
ICT is the Image Compression Technique used for image processing,
BD is Block Dependency method used for image processing,
IPM is Image Partitioning Method used for image processing,
MA is Memory Architecture of the parallel processor used,
MO/NT is Memory Organisation / Network Topology of the parallel processor used,
NP is Number of Processors used.

The options for each of the parameters listed above were described in section 3.3. A cross product of all parameter options will give a very large set of Plans. In this project a total of fifteen Plans were modelled. These Plans are divided into two set: a set of twelve Plans for Non-Inter-Processor Communication (NIPC) method and a set of three Plans for Inter-Processor Communication (IPC) method; as described in the following sections.

5.3.1.1 Plans for Non-Inter-Processor Communication (NIPC) Method: The JPEG algorithm is called as the Non-Block Comparator Technique (NBCT) and the proposed enhancement to the JPEG algorithm is called as the Block Comparator Technique (BCT) in this book. A total of twelve Plans were selected for modelling and simulation of image compression for the Non-Inter-Processor Communication method. These Plans can be classified as Plans for the Non-Block Comparator

Technique (NBCT) and Plans for the Block Comparator Technique (BCT). These Plans are:

Selected Plans using the Non-Block Comparator Technique;

P2	=	P(NBCT, NIPC, BWIP, SMA, SGM, NP),
P3	=	P(NBCT, NIPC, BWIP, SMA, SLgM, NP),
P4	=	P(NBCT, NIPC, BWIP, DMA, DTrT, NP),
P5	=	P(NBCT, NIPC, BWIP, DMA, DToT, NP),
P6	=	P(NBCT, NIPC, BWIP, DMA, DPyT, NP),
P7	=	P(NBCT, NIPC, BWIP, DMA, DCuT, NP),

and selected Plans using the Block Comparator Technique;

P1	=	P(BCT, NIPC, BWIP, DMA, DTrT, NP),
P8	=	P(BCT, NIPC, BWIP, SMA, SGM, NP),
P9	=	P(BCT, NIPC, BWIP, SMA, SLgM, NP),
P10	=	P(BCT, NIPC, BWIP, DMA,DToT, NP),
P11	=	P(BCT, NIPC, BWIP, DMA, DPyT, NP),
P12	=	P(BCT, NIPC, BWIP, DMA, DCuT, NP).

In these Plans the Non-Inter-Processor Communication (NIPC) method was selected for the Block Dependency parameter and Balanced Workload Image Partitioning (BWIP) method for Image Partitioning Method. All of these Plans are simulated for various number of processors.

The example Plan P1 given in equation 5.1 is based on the Non-Inter-Processor Communication method, because the neighbouring samples are transmitted along with the main image block. This can be seen in the **Send ImgPart** instruction on the Host Processor. If the Inter-Processor Communication method is used for transferring neighbouring samples the communication time leads to increase in total execution time compared to that for the Non-Inter-Processor Communication method. Therefore, the Non-Inter-Processor Communication (NIPC) method was chosen.

Block Comparison using Divide and Conquer sort method was used for the Block Comparator Technique. This method is also convenient for grouping the blocks that have equal intensity values, for the Balanced Workload Image Partitioning (BWIP) method. In the Balanced Workload Image Partitioning (BWIP), time for which processors are idle can be minimised. Therefore, the Balanced Workload Image Partitioning method was selected for the simulation.

The total number of simulation models developed for the Non-Inter-Processor Communication method are as follows,

Total Number of Models for the NIPC Plans

= Number of Models for the NBCT + Number of Models for the BCT.

Number of Models for NBCT

= Number of Models for Shared Memory Architecture + Number of Models for Distributed Memory Architecture

= NMSMG + NMSMGL + NMDMTr + NMDMTo + NMDMPy + NMDMCu

≅ 9 x 3 + 8 x 3 + 8 x 3 + 8 x 3 + 6 x 3 + 5 x 3 ≅ 134 Models,

where,

NMSMG	=	Number of Models for Shared Memory Architecture with Global Memory organisation,
NMSMGL	=	Number of Models for Shared Memory Architecture with Local-plus-Global Memory organisation,
NMDMTr	=	Number of Models for Distributed Memory Architecture with Tree Topology,
NMDMTo	=	Number of Models for Distributed Memory Architecture with Torus Topology,
NMDMPy	=	Number of Models for Distributed Memory Architecture with Pyramid Topology,
NMDMCu	=	Number of Models for Distributed Memory Architecture with Cube Topology.

Number of Models for the Block Comparator Technique (BCT) is almost equal to the number of Models for the Non-Block Comparator Technique (NBCT). Therefore nearly two hundred and sixty eight Models were simulated, and the results obtained from these models are given in the section 5.3.2.

5.3.1.2 Plans for the Inter-Processor Communication (IPC) Method: A total of three Plans were selected for modelling and simulation of image compression for the Inter-Processor Communication (IPC) method. The representation of these Plans for the Block Comparator Technique are:

Selected Plans for the Block Comparator Technique with the Inter-Processor Communication (IPC) method;

P13 = P(BCT, IPC, BWIP, SMA, SlgM, NP),
P14 = P(BCT, IPC, BWIP, DMA, DToT, NP),
P15 = P(BCT, IPC, BWIP, DMA, DPyT, NP).

For these Plans the Block Comparator Technique (BCT) was selected for Image Compression Technique parameter. For the Inter-Processor Communication method, Plans for Shared Memory Architecture with Global Memory organisation and Distributed Memory Architecture with Torus and Pyramid topologies were simulated. This helps to compare the execution times with NIPC Plans.

The total number of simulation models developed for the Inter-Processor Communication (IPC) method are as follows:

Total Number of Models for the IPC Plans

= Number of Models for Shared Memory Architecture + Number of Models for Distributed Memory Architecture

= NMSMGL + NMDMTo + NMDMPy

≅ 9 x 3 + 8 x 3 + 8 x 3 ≅ 75 Models,

where,

NMSMGL	=	Number of Models for Shared Memory Architecture with Local-plus-Global Memory organisation,
NMDMTo	=	Number of Models for Distributed Memory Architecture with Torus Topology,
NMDMPy	=	Number of Models for Distributed Memory Architecture with Pyramid Topology.

Therefore, nearly seventy five models were simulated for the IPC Plans, and results obtained from these models are given in the section 5.3.2.

5.3.2 Execution Times Obtained

The execution times obtained from simulation experiments are given in appendix A. Tables A.1 to A.15 show the execution times for the Plans mentioned above for various number of processors and for three different image sizes. The image sizes considered are 125 x 125 samples, 625 x 425 samples and 1100 x 900 samples; as these are the most widely used image sizes in the industry. From all the tables it can be seen that the execution time increases with the increase in image size, and the execution time decreases with the increase in Number of Processors (NP) till some point, then it starts increasing with NP. The point at which the least execution time is obtained gives the scaleup for that specific Plan.

Execution times for the Non-Block Comparator (NBCT) using the Non-Inter-Processor Communication (NIPC) method are given in tables A.1 to A.6. The execution times for the Block Comparator Technique (BCT) using the NIPC method are given in tables A.7 to A.12. The Number of Similar Blocks (NSB) is 10% for tables A.7 to A.12 NSB = 10%. The execution times for the Block Comparator Technique (BCT) using Inter-Processor Communication (IPC) method are given in tables A.13 to A.15; in these tables also the Number of Similar Blocks (NSB) is 10 %. The least execution time in all these tables are indicated in bold letters.

Comparison of execution times obtained for the NIPC Plans is given in section 5.3.2.1. Execution time comparison for the IPC Plans is given in section 5.3.2.2. Execution time comparison between the NIPC and the IPC Plans is given in section 5.3.2.3.

5.3.2.1 Comparison of Execution Times for the NIPC Plans: In this section a comparison of execution times for the Non-Inter-Processor Communication (NIPC) Plans is given. The execution times are tabulated in tables A.1 to A.6. From these tables the least execution time values obtained for the different Plans are extracted, tabulated and compared in this section.

A comparison of execution times obtained from simulation is more meaningful than a comparison of execution times obtained on real systems, such as the three systems discussed in chapter 4. Because the processing power of the CPUs used on different real systems can be quite different. Whereas, the processing power of the different CPUs used in simulation are the same. The simulation models discussed in this book are based on the Intel i860 CPU.

Section 5.3.2.1a gives the execution times comparison for the NBCT Plans, and execution times comparison for the BCT Plans is given in section 5.3.2.1b. Comparison of these two techniques, based on the Speed Improvement Factor is given in section 5.3.2.1c.

5.3.2.1a Comparison of Execution Times for the NBCT Plans: Table 5.2 gives the least execution times extracted from the tables A.1 to A.6, which are based on the NBCT Plans. By comparing the least execution times obtained on Plans P2 and P3, it can be seen that the least execution time on Shared Memory Architecture with Local-plus-Global Memory organisation is lower.

By comparing Plans for Distributed Memory Architecture, we can see that the least execution time for the 125 x 125 image size on Plan P5 is the lowest, and for the other image sizes the least execution time for P6 is the lowest. Therefore, we can say that the least execution time can be obtained for small image size on Distributed Memory with Torus Topology and for medium and large image sizes lowest execution time can be obtained on Distributed Memory Architecture with Pyramid Topology.

By comparing all Plans we can see that the least execution time for the Non-Block Comparator Technique (NBCT) can be obtained on the Shared Memory Architecture with Local-plus-Global Memory organisation with twenty processors.

Table 5.2 Least execution times for the NBCT Plans

Plan	Memory Architecture	Memory organisation / topology	Execution times in msec.					
			NP	For 125 x 125 image	NP	For 625 x 429 image	NP	For 1100 x 900 image
P2	Shared Memory	Global	16	17.82	25	235.49	25	869.13
P3	Architecture	Global & Local	**20**	**10.47**	**20**	**194.96**	**20**	**732.21**
P4	Distributed	Tree	27	15.49	27	267.72	27	981.38
P5	Memory	Torus	**21**	**11.69**	21	262.58	21	959.29
P6	Architecture	Pyramid	37	13.43	**37**	**224.55**	**37**	**812.04**
P7		Cube	28	16.30	28	277.49	28	959.14

5.3.2.1b Comparison of Execution Times for the BCT Plans: Table 5.3 gives the least execution times for all the Plans based on the BCT. By comparing the least execution times for P8 and P9 Plans, we can see that the least execution time on the Shared Memory Architecture with Local-plus-Global Memory organisation is lower as compared to that on the Shared Memory Architecture with Global Memory organisation.

By comparing the Plans for Distributed Memory Architecture, we can see that the least execution time for all image sizes is the lowest on Plan P11. Therefore, we can say that the least execution time can be obtained on Distributed Memory Architecture with Pyramid Topology.

Table 5.3 Least execution times for selected Plans using the Block Comparator Technique

Plan	Memory Architecture	Memory organisation /topology	Execution times in msec.					
			NP	For 125 x 125 image	NP	For 625 x 429 image	NP	1100 x 900 image
P2	Shared Memory	Global	16	16.62	16	275.64	16	1005.1
P3	Architecture	Global & Local	**20**	**10.73**	**20**	**200.42**	**20**	**733.24**
P4	Distributed	Tree	15	15.96	15	259.58	15	1030.6
P5	Memory	Torus	9	21.33	9	348.42	13	1173.7
P6	Architecture	Pyramid	**21**	**13.59**	**21**	**226.41**	**21**	**833.01**
P7		Cube	9	19.26	9	318.53	9	1165.6

By comparing all Plans we can see that the least execution time for the Block Comparator Technique (BCT) can be obtained on Shared Memory Architecture with Local-plus-Global Memory organisation.

5.3.2.1c Speed Improvement Factor: This section presents a comparison of the BCT and the NBCT Plans based on the Speed Improvement Factor. The speed improvement obtained by the Block Comparator Technique over the Non-Block Comparator Technique can be represented by a factor called the Speed Improvement Factor (SIF), defined in chapter 2. SIF is defined as the ratio of execution time obtained for the Non-Block Comparator Technique (NBCT) to the execution time obtained for the Block Comparator Technique (BCT), as given by

$$SIF = \frac{T_{NBCT}}{T_{BCT}}$$

SIF values for Shared Memory Architecture with Global Memory organisation are obtained by comparing the execution times obtained for Plans P2 and P8; these are given in table 5.4, and plotted in figure 5.3 for NSB = 10%.

Table 5.4 SIF values for the NBCT Plan P2 and the BCT Plan P8
(on a Shared Memory Architecture with Global Memory organisation)

Number of Processors-NP	Speed Improvement Factor - SIF		
	For 125 x 125 image	**For 625 x 429 image**	**For 1100 x 900 image**
1	1.07	1.07	1.07
3	1.03	1.07	1.17
5	1.11	1.05	1.06
7	1.06	1.00	1.06
11	1.11	**1.02**	1.03
16	**1.07**	0.97	**1.01**
20	0.99	0.83	0.84

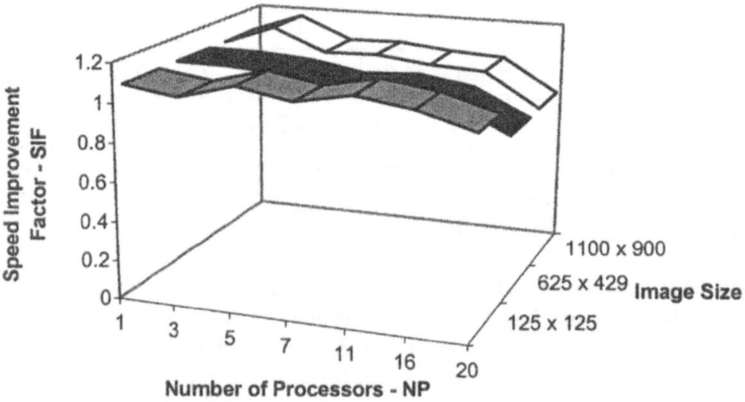

Figure 5.3 SIF graph for Plans P2 and P8

The SIF values of 1.07 obtained from simulation, for NP = 1, is close to the SIF value of 1.08 obtained by analytical means; given in table 2.8. From table 5.4 it can be seen that SIF value is greater than one upto some Number of Processors and then it falls below one. The highest Number of Processors at which SIF is greater than one is called as the Speed Improvement Cutoff Point (SICP) in this book. In appendix A tables A.16 to A.20 give the SIF values for the various architectures. From these tables the SICP values are extracted and tabulated in table 5.5.

Table 5.5 SIF values for various Plans

Memory Architecture	Memory organisation / topology	Speed Improvement Cutoff Point (SICP)		
		For 125 x 125 image	For 625 x 429 image	For 1000 x 900 image
Shared Memory Architecture	Global	**16**	**11**	**16**
	Global & Local	7	11	11
Distributed Memory Architecture	Tree	15	15	15
	Torus	5	5	13
	Pyramid	**21**	**21**	**21**
	Cube	9	9	9

From the table 5.5 we can see that the highest value of SICP is obtained for the Pyramid Topology. Of the two Shared Memory Architectures the Global Memory organisation gives a higher value for the SICP. From the above discussion it can be concluded that the Block Comparator Technique fails to be effective on a large Number of Processors. The reason for this is explained below.

In this chapter the divide and conquer method was used, as outlined in the chapter 2, for comparing blocks in an image. In all of the techniques discussed in this chapter the block comparison step takes place on the host processor and leads to sequential processing of the block comparison process. This leads to the fact that the time taken for the Block Comparator Technique is higher than that for the Non Block Comparator Technique as the number of processors increases. Thus execution time can be further reduced by parallelising the block comparison step as well [Kumar, 94] [Zomaya, 96].

5.3.2.2 Comparison of Execution Times for the IPC Plans : The execution times obtained for the IPC Plans P13 to P15 are given in the appendix tables A.13 to A.15 respectively. The least execution times for the IPC Plans are given in table 5.6. By comparing the three Inter-Processor Communication based Plans, we can observe that the Distributed Memory Architecture with Pyramid Topology gives the least execution time.

The execution time obtained on the Shared Memory Architecture with Local-plus-Global Memory organisation is higher due to contention over the transfer device. In the Distributed Memory Architecture communication takes place over the various transfer devices in parallel. This can be seen by comparing utilisation graphs of Transfer Devices given in figures 5.16 and 5.17.

Table 5.6 Least execution times for the IPC Plans Plans using the Block Comparator Technique

Plan	Memory Architecture	Memory organisation / topology	Execution times in msec.					
			NP	For 125 x 125 image	NP	For 625 x 429 image	NP	For 1100 x 900 image
P13	Shared Memory	Global & Local	13	20.09	13	299.42	13	1092.95
	Architecture							
P14	Distributed	Torus	13	18.80	13	295.36	13	1069.1
P15	Memory	Pyramid	**21**	**12.03**	**21**	**220.21**	**21**	**830.27**
	Architecture							

5.3.2.3 Comparison of Execution Times for Different Block Dependency Method (NIPC and IPC): By comparing Plans for the Non-Inter-Processor Communication (NIPC) method and the Inter-Processor Communication (IPC) method using the Block Comparator Technique, the following conclusions are derived.

Comparing Plan P9 and P13: Execution times for Plan P13 (BCT on a Shared Memory Architecture with Global Memory organisation for Inter Processor Communication) are higher than that for Plan P9 (BCT on a Shared Memory Architecture with Local-plus-Global Memory organisation for Non-Inter Processor Communication), because of memory contention problem on a shared bus.

Comparing Plans P10 and P14, Plans P11 and P15: The execution times for Plan P14 (BCT on a Distributed Memory Architecture with Torus Topology for the IPC) and P15 (BCT for Distributed Memory Architecture with Pyramid Topology for the IPC) are higher than those for Plans P10 (BCT on a Distributed Memory Architecture with Torus Topology for the NIPC) and P11 (BCT on a Distributed Memory Architecture with Pyramid Topology for the NIPC) respectively for Number of Processors ≤ 5. For Number of Processors > 5 the execution times for Plans P14 and P15 are lower than those for Plans P10 and P11 respectively.

From this we can conclude that the IPC is efficient for lower Number of Processors. Because the communication time for lower Number of Processors for inter-communication among processors is less as compared to the NIPC. As the increase in NP, the communication time taken for transferring neighbouring samples among processors at run time increases. This leads to increase in execution time.

5.3.2.4 Comparison of Execution Times for Plan P11 with Different NSB : In the BCT Plans discussed in section 5.3.2.1, the Number of Similar Blocks (NSB) in an image was taken as 10%. From the execution times comparison of these Plans, the Pyramid topology was found to be the best. Therefore, to compare the execution times for different values of NSB, the Pyramid Topology is selected for simulation.

The execution times obtained on the Pyramid Topology architecture for the NBCT Plan P6 and sequential processing of block comparison in Plan P11 for 625 x 423 image size, using NSB = 10%, 30%, 50% and 75% are given in table 5.7. SIF values derived from the data given in table 5.7 are given in table 5.8. From the execution times table it can be seen that the least execution time is obtained on 21 processors. SIF values shows that the maximum SIF value can be obtained on 21 processors. After that the SIF value decreases, because the communication time is considerably higher as compared to NBCT on higher Number of Processors. From table 5.8, we can see that the SIF value of 1.07 for NSB = 10%, 1.38 for NSB = 30%, 1.86 for NSB = 50%, and 3.44 for NSB = 75% obtained from simulation, for NP = 1, is close to the SIF value of 1.08, 1.377, 1.889, 3.526 respectively obtained by analytical means; given in table 2.8.

Table 5.7 Ex. times for NBCT Plan P6 and sequential block comparison with Plan P11

Number of Processors-NP	**Execution times in msec.**				
	NBCT	**10% NSB**	**30% NSB**	**50% NSB**	**75% NSB**
1	1871.8	1745.6	1356.37	1006.3	544.14
5	386.07	373.23	283.17	213.66	116.02
9	359.84	318.53	240.92	181.32	100.77
21	309.03	226.41	186.31	145.43	79.21
37	224.55	239.29	196.86	176.55	103.45

Table 5.8 SIF values for Plan P6 and Plan P11

Number of Processors-NP	**Speed Improvement Factor - SIF**			
	10% NSB	**30% NSB**	**50% NSB**	**75% NSB**
1	1.07	1.38	1.86	3.44
5	1.03	1.36	1.81	3.33
9	1.13	1.49	1.99	3.57
21	**1.36**	**1.66**	**2.12**	**3.90**
37	0.94	1.14	1.27	2.17

5.4 Performance Comparison of Parallel Architectures

Performance of parallel computer architectures can be measured in terms of speedup, scaleup and efficiency. Tables A.21 to A.35 give the speedup for the simulated Plans for various Number of Processors and for three different image sizes and speedup graph of the same are shown in figure A.1a to A.15a respectively. From these tables and graphs we can see that the speedup increases with the increase in Number of Processors (NP) and at some point it starts decreasing with the increase in Number of Processors. The point at which the speedup starts decreasing is defined as the scaleup for the Plan.

Speedups for the Non-Block Comparator Technique Plans with Non-Inter-Processor Communication (NIPC) are given in tables A.21 to A.26. Tables A.27 to A.32 gives the speedup for the Block Comparator Technique Plans with Non-Inter-Processor Communication (NIPC) in which Number of Similar Blocks (NSB) = 10%. Speedup obtained for the Inter-Processor Communication Plans for the Block Comparator Technique is given in tables A.33 to A.35. Comparison of speedup values obtained for different Plans are discussed in the following sections.

5.4.1 Comparison of Speedup

Speedup for NP processors was defined in chapter 3, as the time taken by a single processor divided by the time taken by NP processors. Speedup (S) of parallel processor is given by,

$$S = T_1 / T_{NP}$$

where,

S = Speedup for N processors,

T_1 = Time taken by a single processor,

T_{NP} = Time taken by NP processors.

Comparison of speedup obtained for the NIPC Plans is given in section 5.4.1.1. Comparison of speedup obtained for the IPC Plans is given in section 5.4.1.2. Section 5.4.1.3 gives the comparison of speedup obtained for different values of NSB.

5.4.1.1 Comparison of Speedup for the NIPC Plans: This section gives the comparison of speedup for the NIPC Plans. Comparison of maximum speedup for the NBCT Plans is discussed in section 5.4.1.1a and that for the BCT Plans is discussed in section 5.4.1.1b. Comparison of maximum speedup of NBCT and BCT Plans is given in section 5.4.1.1c.

5.4.1.1a Comparison of Maximum Speedup for the NBCT Plans: Table 5.9 shows a comparison of the maximum speedup for the various NBCT Plans, which are extracted from the appendix A; tables A.21 to A.26. From these tables the following conclusions can be derived:

1. For the Non-Block Comparator Technique the higher values of speedup can be obtained on the Shared Memory Architecture with Local-plus-Global Memory organisation.
2. Of the various Distributed Memory Architectures, highest values of speedup can be obtained on the Torus Topology for the 125 x 125 image size, the Pyramid Topology gives higher values of speedup for the other two image sizes.
3. Of the two different Shared Memory Architectures, higher value of speedup can be obtained on the Shared Memory Architecture with Local-plus-Global Memory organisation for all image sizes.

Table 5.9 Maximum speedup comparison for the NBCT Plans

Plan	Memory Architecture	Memory organisation / topology	Speedup-S					
			NP	125 x 125 image	NP	625 x 429 image	NP	For 1100 x 900 image
P2	Shared Memory	Global	16	6.30	25	7.95	25	7.87
P3	Architecture	Global & Local	**20**	**10.73**	**20**	**9.60**	**20**	**9.35**
P4	Distributed	Tree	27	7.25	27	6.99	27	6.97
P5	Memory	Torus	**21**	**9.61**	21	7.13	21	7.13
P6	Architecture	Pyramid	37	8.36	**37**	**8.34**	**37**	**8.42**
P7		Cube	28	6.89	28	6.75	28	7.14

5.4.1.1b Comparison of Maximum Speedup for the BCT Plans: Table 5.10 shows a comparison of the maximum speedup values obtained for the various BCT Plans, which are extracted from the appendix tables A.27 to A.32. From this table we can arrive at the following conclusions:

1. Of the two Shared Memory Architectures, the higher values of speedup can be obtained on the Shared Memory Architecture with Local-plus-Global Memory organisation.
2. Of the various Distributed Memory Architectures, highest value of speedup can be obtained on the Pyramid Topology for all image sizes.
3. Of the two different Architectures for the Non-Block Comparator Technique the higher values of speedup can be obtained on the Shared Memory Architecture with Local-plus-Global Memory organisation.

Table 5.10 Maximum speedup comparison for the BCT Plans

Plan	Memory Architecture	Memory organisation / topology	Speedup-S					
			NP	125 x 125 image	NP	625 x 429 image	NP	For 1100 x 900 image
P8	Shared Memory	Global	16	6.30	16	6.33	16	6.35
P9	Architecture	Global & Local	**20**	**9.75**	**20**	**8.71**	**20**	**8.70**
P1	Distributed	Tree	15	6.56	15	6.72	15	6.19
P10	Memory	Torus	9	4.91	9	5.01	13	5.44
P11	Architecture	Pyramid	**21**	**7.70**	**21**	**7.71**	**21**	**7.66**
P12		Cube	9	5.43	9	5.48	9	5.48

5.4.1.1c Comparison of Speedup for the NBCT and BCT Plans: Table 5.11 gives a comparison of the maximum speedups obtained for the NBCT Plan and the BCT Plans. From this table we can see that the maximum speedups for the NBCT Plan are higher as compared to the same for the BCT Plans.

Table 5.11 Maximum speedup comparison for the NBCT and the BCT Plans

Memory Architecture	Memory organisation / topology	Speedup-S					
		For 125 x 125 image		For 625 x 429 image		For 1100 x 900 image	
		NBCT	BCT	NBCT	BCT	NBCT	BCT
Shared Memory	Global	6.30	6.30	7.95	6.33	7.87	6.35
Architecture	Global & Local	**10.73**	**9.75**	**9.60**	**8.71**	**9.35**	**8.70**
Distributed	Tree	7.25	6.56	6.99	6.72	6.97	6.19
Memory	Torus	**9.61**	4.91	7.13	5.01	7.13	5.44
Architecture	Pyramid	8.36	**7.70**	**8.34**	**7.71**	**8.42**	**7.66**
	Cube	6.89	5.43	6.75	5.48	7.14	5.48

5.4.1.2 Comparison of Speedup for the IPC Plans: Table 5.12 gives the speedup comparison of the Shared Memory and the Distributed Memory Architecture. Comparing these two architectures we can see that,

- The speedup values obtained for the Pyramid Topology (Plan 15) is higher than that for the other two.

Table 5.12 Speedup comparison of two architectures

Plan	Memory Architecture	Memory organisation / topology	Speedup-S					
			NP	125 x 125 image	NP	625 x 429 image	NP	For 1100 x 900 image
P13	Shared Memory Architecture	Global & Local	13	5.21	13	5.83	13	5.84
P14	Distributed Memory Architecture	Torus	13	5.57	13	5.91	13	5.97
P15		Pyramid	**21**	**8.70**	**21**	**7.93**	**21**	**7.69**

5.4.1.3 Comparison of Speedup for Different NSB: Table 5.13 gives the speedup on the Pyramid Topology with 625 x 429 image size for NSB = 10%, 30%, 50%, and 75% and speedup graphs of the same are shown in figure 5.4. From this table the following conclusions can be derived:

- Speedup values for the BCT are higher those for the NBCT for NP ≤ 21. For NP = 37 the NBCT gives higher speedup as compared to the BCT.
- By comparing the speedup values obtained for NSB = 10% and 30%, it can be seen that speedup for NSB = 30% is higher than that of for NSB = 10% for NP ≤ 9.
- By comparing the speedup values obtained for NSB = 30%, 50% and 75%, it can be seen that the speedup decreases as NSB increases for all values of NP. This indicates that as NSB increases, the computation time for block comparison is significantly higher than the computation time for comparing the image on the parallel processor.

Table 5.13 Speedup for the NBCT Plan P6 and the BCT Plan P11 with different NSB

Number of Processors-NP	Speedup - S				
	NBCT	NSB=10%	NSB=30%	NSB=50%	NSB=75%
1	1	1	1	1	1
5	4.65	4.68	4.79	4.71	4.69
9	5.2	5.48	5.63	5.55	5.4
21	6.06	**7.71**	**7.28**	**6.92**	**6.87**
37	8.34	7.29	6.89	5.7	5.26

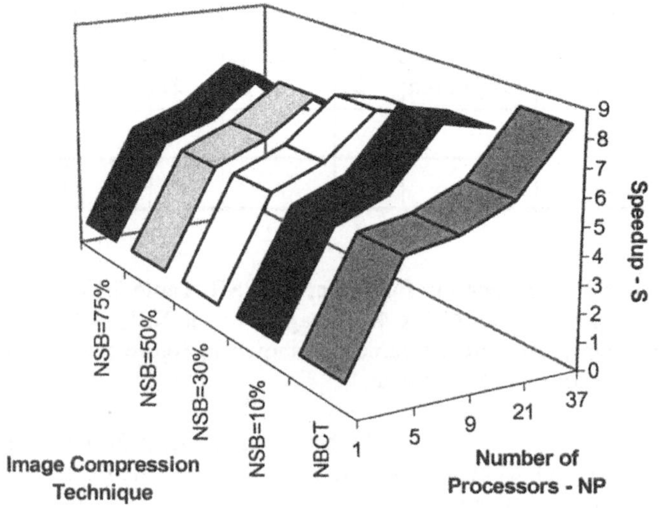

Figure 5.4 Speedup graph for Plans P6 and P11 for different NSB values

5.4.2 Comparison of Scaleup

Scaleup of a parallel architecture is a function of the maximum number of processors at which the speedup starts decreasing. Scaleup was defined in chapter 3, as the number of processors at the point of maximum speedup. Section 5.4.2.1 gives a comparison of the scaleup values obtained for the NIPC Plans and section 5.4.2.2 gives a comparison of scaleup values obtained for the IPC Plans.

5.4.2.1 Scaleup Comparison for the NIPC Plans: A comparison of scaleup obtained for the Non-Block Comparator Technique (NBCT) is given in section 5.4.2.1a and the same for the Block Comparator Technique (BCT) in section 5.4.2.1b. Comparison of the two techniques is given in section 5.4.1.1c.

5.4.2.1a Scaleup Comparison for the NBCT Plans: Table 5.14 shows a comparison of the scaleup obtained for the NBCT Plans. From this table the following conclusions can derived:

1. For the Shared Memory Architectures the hybrid memory organisation (Plan P3) gives higher scaleup than a purely Global Memory organisation for 125 x 125 (small) image size. For the medium and the large image sizes the Global Memory organisation (Plan P2) gives higher values of scaleup.
2. The Distributed Memory Architectures give highest values of scaleup as compared to the Shared Memory Architectures.
3. Pyramid Topology (Plan P6) gives the best values for scaleup.

Table 5.14 Scaleup comparison for the NBCT Plans

Memory Architecture	**Plan**	**Scaleup-S**		
		For 125 x 125 image	**For 625 x 429 image**	**For 1100 x 900 image**
Shared Memory	P2	16	**25**	**25**
Architecture	P3	**20**	20	20
Distributed Memory	P4	27	27	27
Architecture	P5	21	21	21
	P6	**37**	**37**	**37**
	P7	28	28	28

5.4.2.1b Scaleup Comparison of the BCT Plans: Table 5.15 shows a comparison of the scaleup values obtained for the BCT Plans. From this table the following conclusions can be derived:

1. The hybrid memory organisation (Plan P9) gives higher scaleup than a purely Global Memory organisation for all image sizes.
2. The Distributed Memory Architectures give higher values of scaleup as compared to the Shared Memory Architectures.
3. Pyramid topology (Plan P11) gives the best values for scaleup.

Table 5.15 Scaleup comparison for the BCT Plans

Memory Architecture	Plan	Scaleup-S		
		For 125 x 125 image	For 625 x 429 image	For 1100 x 900 image
Shared Memory	P8	16	16	16
Architecture	P9	**20**	**20**	**20**
Distributed	P1	15	15	15
Memory	P10	9	9	13
Architecture	P11	**21**	**21**	**21**
	P12	9	9	9

5.4.2.1c Scaleup Comparison of the NBCT and the BCT Plans: This section gives a comparison of scaleup values obtained for the BCT and the NBCT Plans. Table 5.16 is derived from the tables 5.14 and 5.15. The following conclusions can be derived from table 5.16:

1. For the Shared Memory Architectures, the scaleup values obtained for the NBCT are either higher or the same as those obtained for the BCT.
2. For the Distributed Memory Architectures, the scaleup values obtained for the NBCT is higher than that obtained for the BCT.

Thus, the Block Comparator Technique does not scaleup as well as the Non-Block Comparator Technique.

Table 5.16 Scaleup comparison for the NBCT and the BCT Plans

Memory Architecture	Memory organisation / topology	Scaleup					
		For 125 x 125 image		For 625 x 429 image		For 1100 x 900 image	
		NBCT	BCT	NBCT	BCT	NBCT	BCT
Shared M	Global	16	16	25	16	25	16
Architecture	Global & Local	**20**	**20**	**20**	**20**	**20**	**20**
Distribute	Tree	27	15	27	15	27	15
Memory	Torus	**21**	9	21	9	21	9
Architect	Pyramid	37	**21**	**37**	**21**	**37**	**21**
ure	Cube	28	9	28	9	28	9

5.4.2.2 Scaleup Comparison of the IPC Plans: Table 5.17 gives a comparison of the scaleup values obtained for the Shared Memory and the Distributed Memory Architectures. By comparing these two architectures it can be seen that,

- Scaleup for the Pyramid Topology is the highest.

Table 5.17 Scaleup comparison of two architectures

Plan	**Memory Architecture**	**Memory organisation / topology**	**Scaleup-S**		
			For 125 x 125 image	**For 625 x 429 image**	**For 1000 x 900 image**
P13	Shared Memory Architecture	Global & Local	13	13	13
P14	Distributed Memory Architecture	Torus	13	13	13
P15		Pyramid	**21**	**21**	**21**

5.4.3 Comparison of Efficiency

Efficiency (η) was defined in chapter 3, as the average speedup of each processor in a parallel processor, and is given by,

$$\eta = S / NP,$$

where:

η = Efficiency,
S = Speedup of the parallel processor,
NP = Number of Processors.

The efficiency graphs for Plans P1 to P15 are shown in appendix A, figures A.1b to A.15b respectively. From these figures we can see that the efficiency of 50% or higher was obtained for some Number of Processors. The highest NP at which efficiency is greater than 50% is called as Efficiency Cutoff Point (ECP) in this book. In other words, from cost-benefit analysis point of view a parallel processor with NP = ECP would be most cost effective.

Section 5.4.2.1 gives the efficiency comparison of the NIPC Plans, and section 5.4.2.2 gives the efficiency comparison of the IPC Plans.

5.4.3.1 Efficiency Comparison of the NIPC Plans: Efficiency graphs for the NBCT Plans are shown in appendix A, figures A.1b to A.6b, and these for the BCT Plans are shown in figures A.7b to A.12b. The ECP values are extracted from these figures and are tabulated in table 5.18. From this table, we can conclude that,

1. Of the different Architectures, the ECP value is the highest on the Shared Memory Architecture with Local-plus-Global Memory organisation.
2. Of the two Shared Memory Architectures, the ECP value is higher on a Shared Memory Architecture with Local-plus-Global Memory organisation.
3. Of the various Distributed Memory Architectures, the ECP value is almost same for all topologies.

Table 5.18 Efficiency Cutoff Point for the NIPC Plans

Memory Architecture	Memory organisation / topology	Efficiency Cutoff Point - ECP					
		For 125 x 125 image		For 625 x 429 image		For 1100 x 900 image	
		NBCT	BCT	NBCT	BCT	NBCT	BCT
Shared Memory Architecture	Global	7	7	11	7	16	7
	Global & Local	**11**	**11**	**11**	**11**	**16**	**11**
Distributed Memory Architecture	Tree	9	9	9	9	9	9
	Torus	**9**	5	9	5	9	5
	Pyramid	9	**9**	**9**	**9**	**9**	**9**
	Cube	9	9	9	9	9	9

5.4.3.2 Efficiency Comparison of the IPC Plans: Efficiency graphs for the BCT Plans are shown in appendix A, figures A.13b to A.15b. The ECP values are extracted from these figures and are tabulated in table 5.19. From this table, we can conclude that,

1. Of the various architectures, the ECP value is higher on Distributed Memory Architectures.
2. Of the two Distributed Memory Architectures, the ECP values is same for Pyramid Topology and Torus Topology.

Table 5.19 Efficiency Cutoff Point for the IPC Plans

Plan	Memory Architecture	Memory organisation / topology	Efficiency Cutoff Point -ECP		
			For 125 x 125 image	For 625 x 429 image	For 1000 x 900 image
P13	Shared Memory Architecture	Global & Local	7	9	9
P14	Distributed Memory	Torus	9	9	9
P15	Architecture	Pyramid	**9**	**9**	**9**

5.5 Summary

Analytical modelling alone is not sufficient to evaluate the behaviour of parallel algorithms. Therefore the Network II.5, a discrete event simulation package, was chosen for the simulation of Digital Image Compression technique.

Simulation of techniques for implementing Digital Image Compression on parallel computers involve developing the models, modelling, validation and experimentation with the models. This chapter explained all of these aspect of the simulation done with the help of the Network II.5 simulation package.

Execution times obtained for the simulation experiments were tabulated for each model. Experimental results confirmed that the Pyramid architecture performed the best in terms of speedup, scaleup and efficiency.

Simulation times obtained for different values of NSB were also tabulated. From speedup figures it could be determined that the scaleup of twentyone and Efficiency Cutoff Point of about nine can be obtained. This implies that even though there is increasing speedup values for upto twenty one processors, the marginal cost of adding more than nine processors is rather high, In other words, from cost-benefit analysis point of view a nine processor system would be most cost effective.

In this chapter the divide and conquer method was used for comparing blocks in an image, and the block comparison step took place on the host processor. Compression time can be further reduced by parallelising the block comparison step as well.

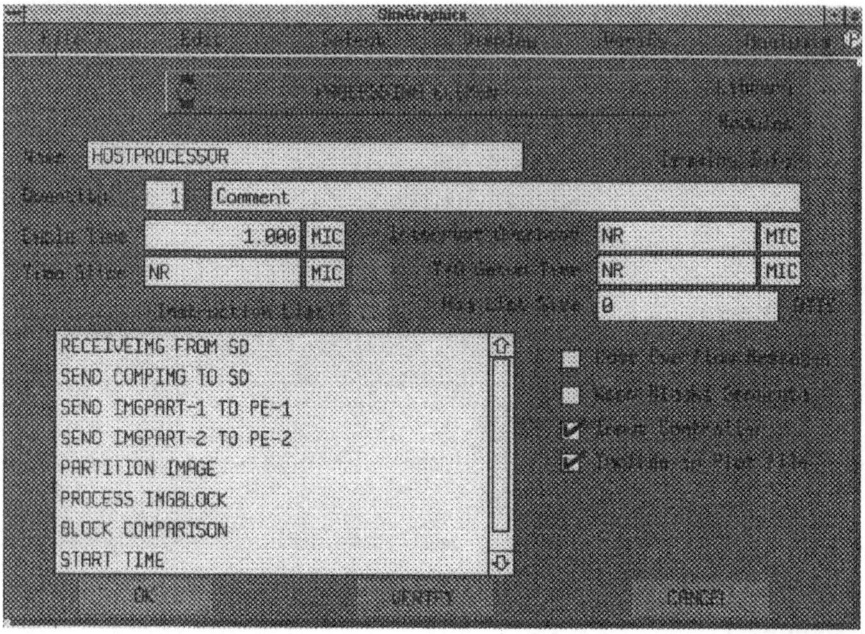

Figure 5.5a Host processor specification form

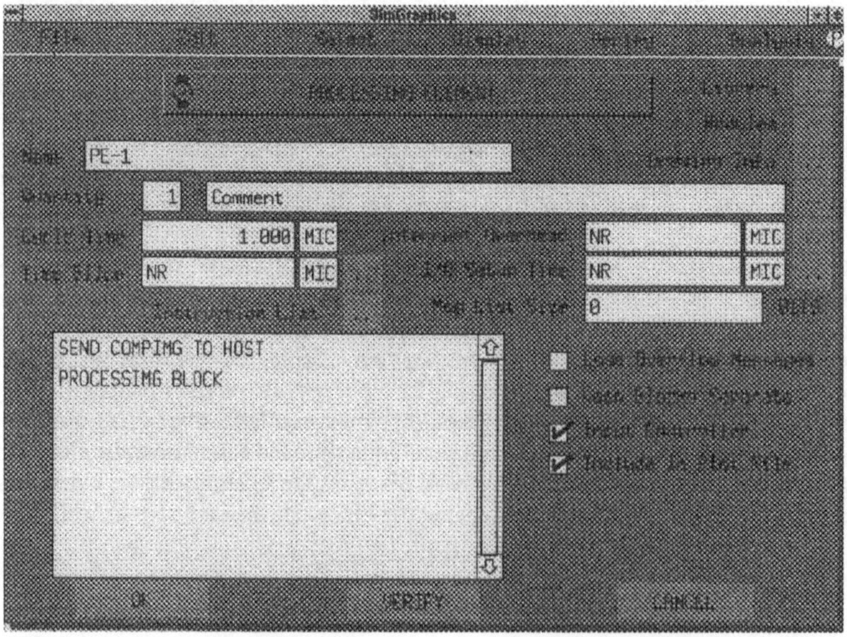

Figure 5.5b PE-1 specification form

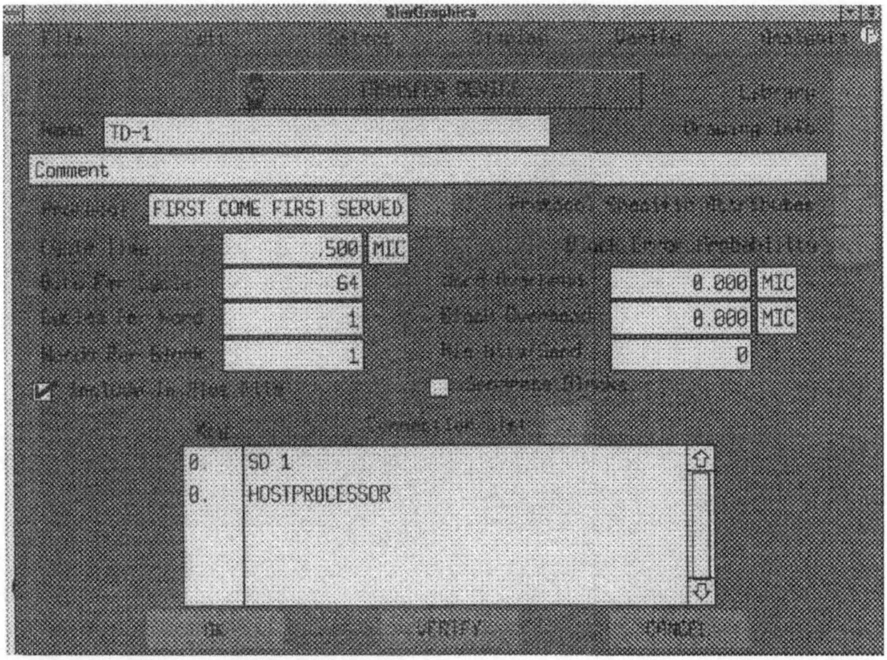

Figure 5.6a TD-1 specification form

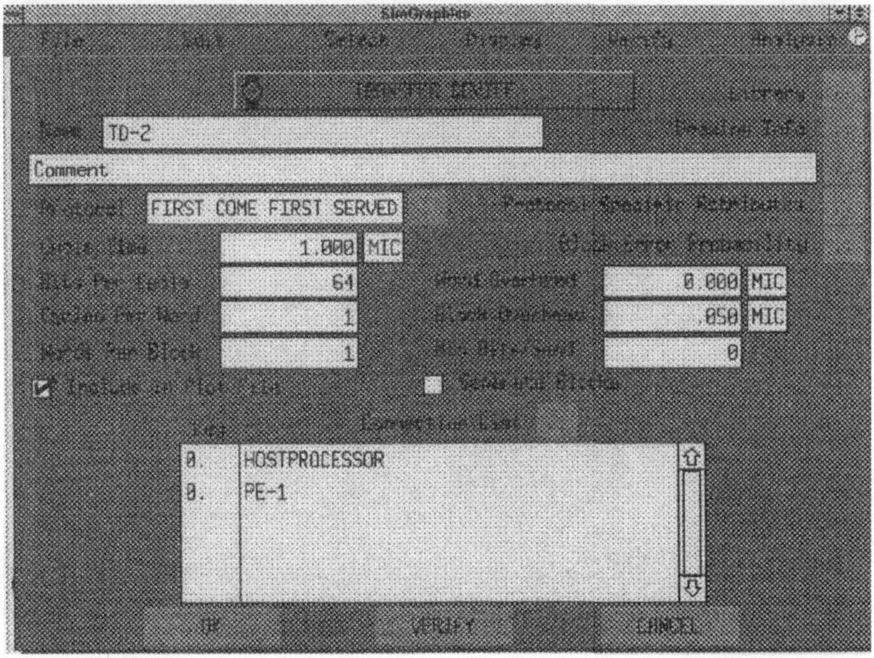

Figure 5.6b TD-2 specification form

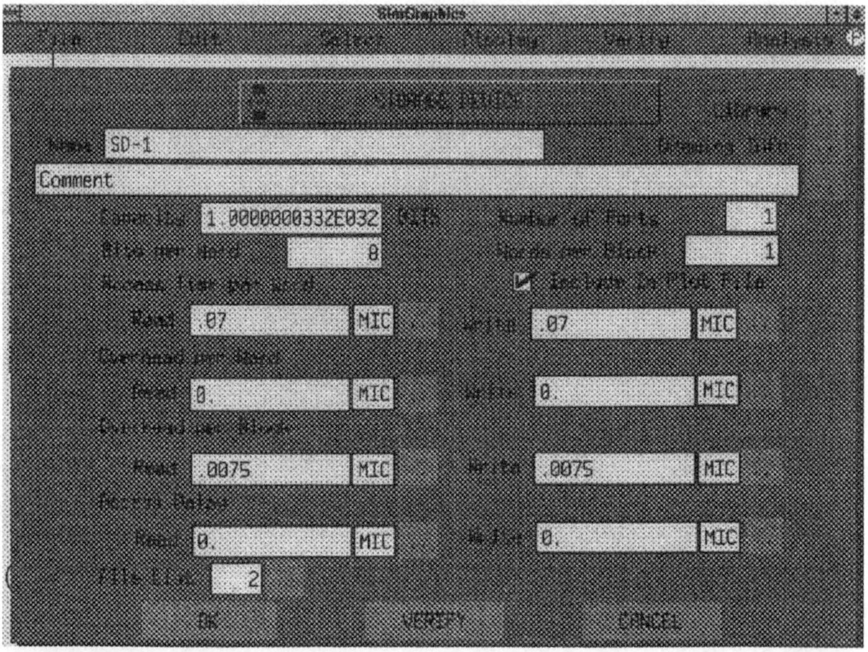

Figure 5.7 SD-1 specification form

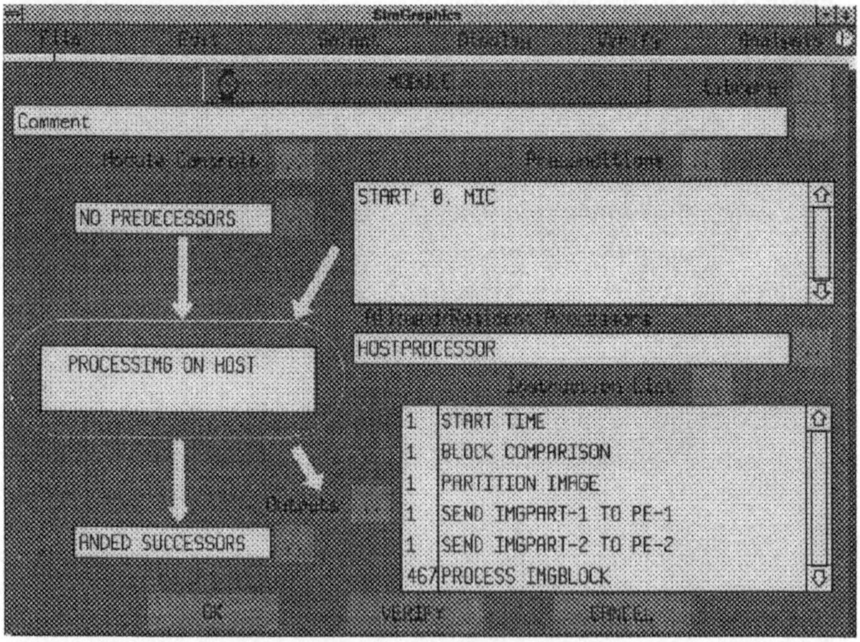

Figure 5.8 Module 1: "ProcessImg on Host"

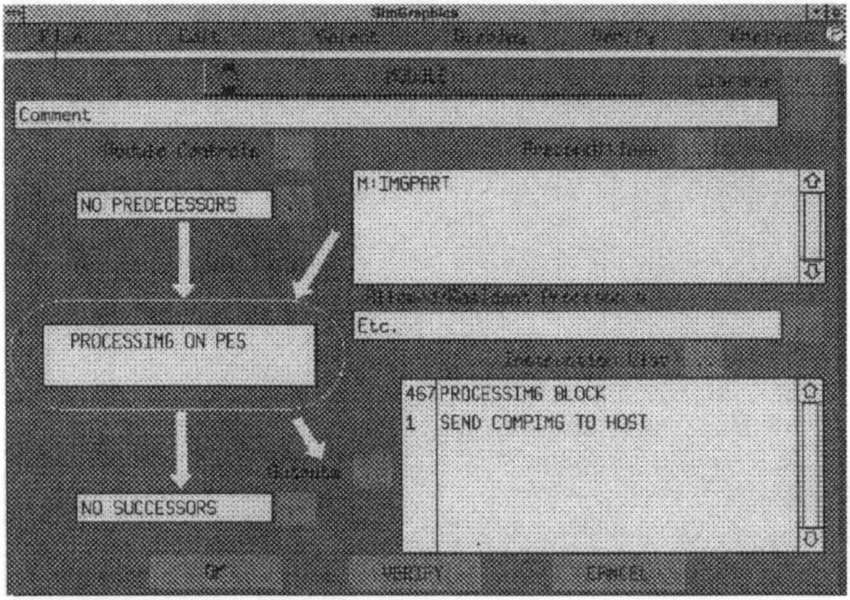

Figure 5.9 Module 2: "ProcessImg on PE"

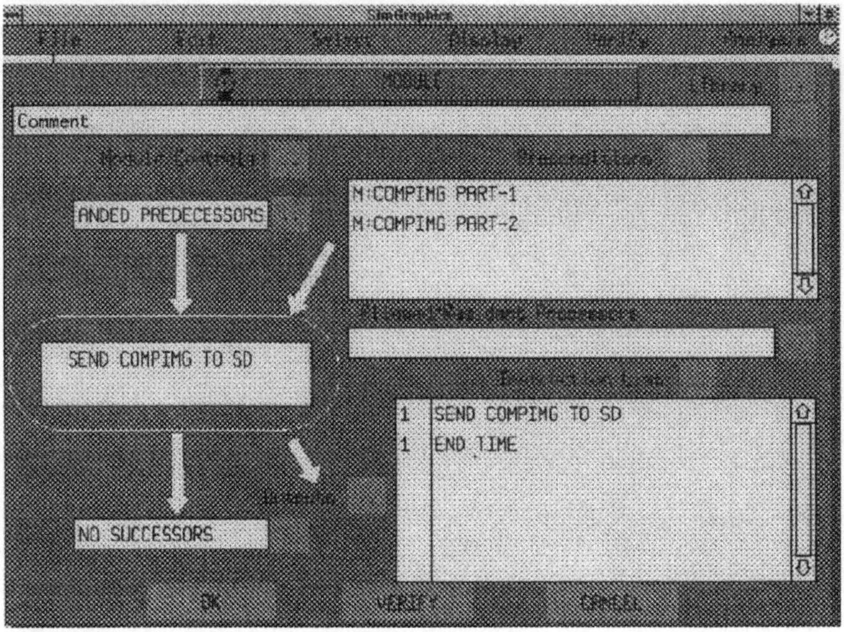

Figure 5.10 Module 3: "Send CompImg to SD"

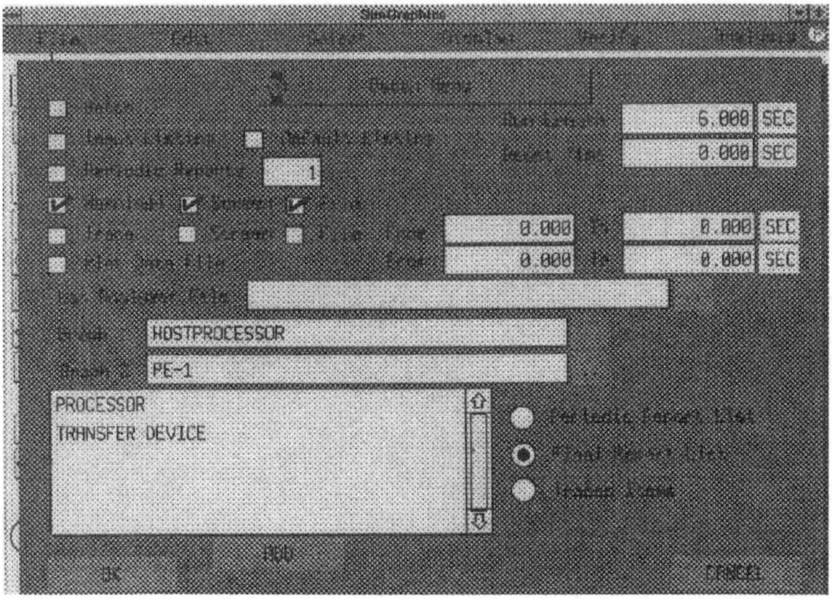

Figure 5.11a Run parameter form

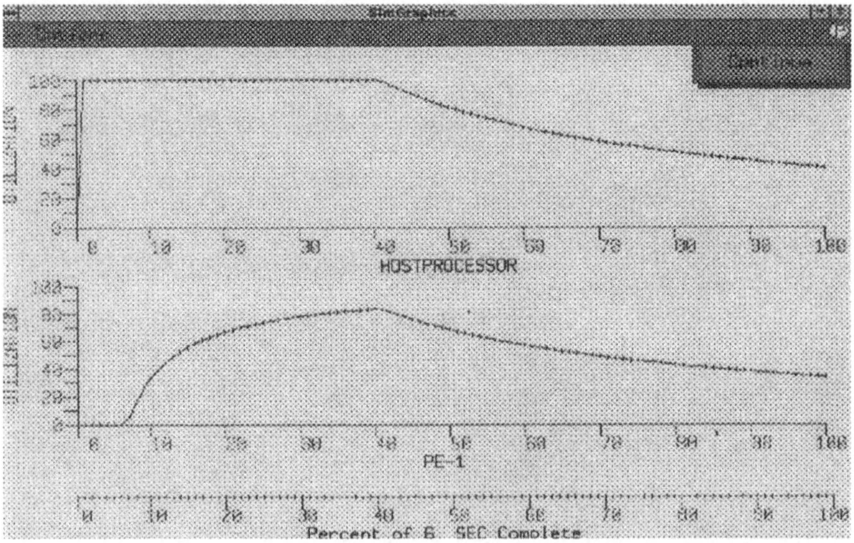

Figure 5.11b Utilisation graph of the host processor and the PE-1 at run time

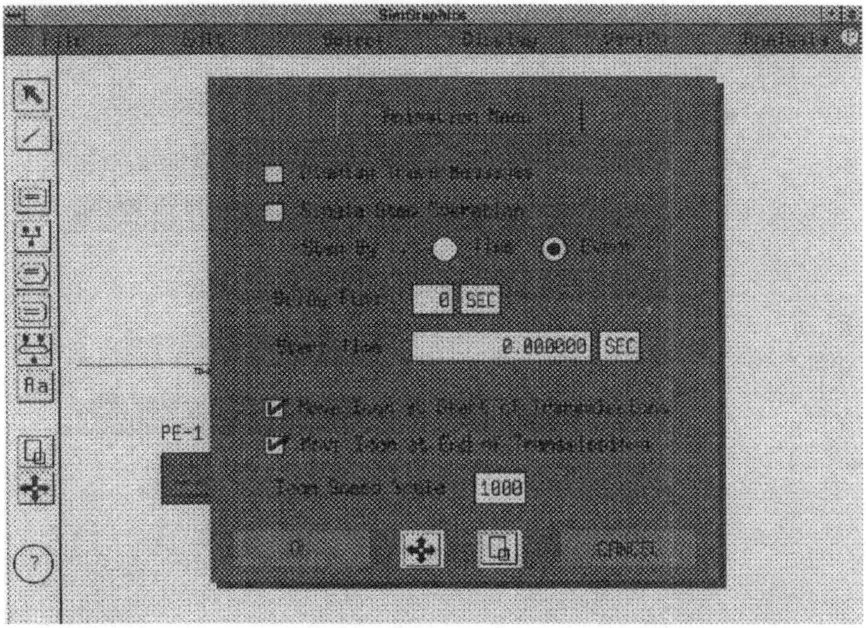

Figure 5.12 Animation parameter specification menu

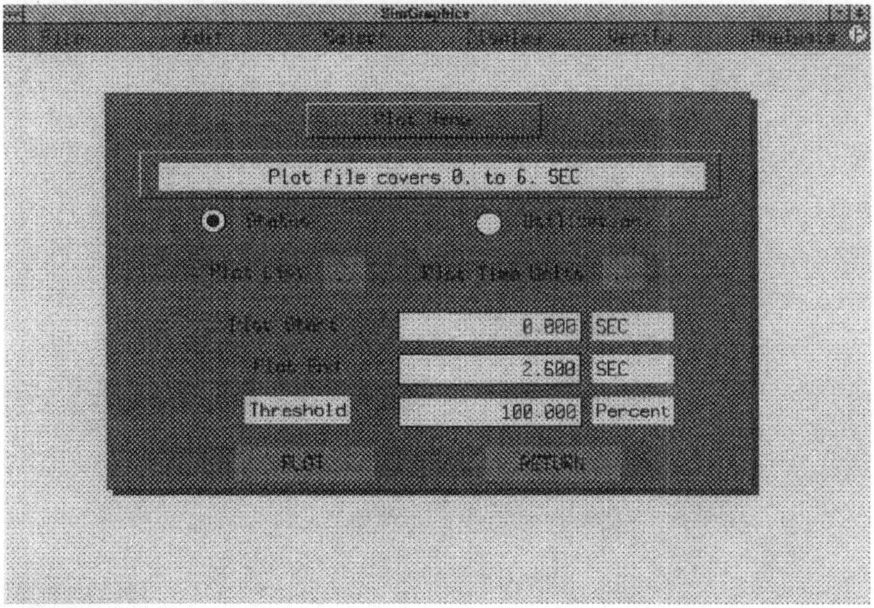

Figure 5.13 Plot parameter specification menu

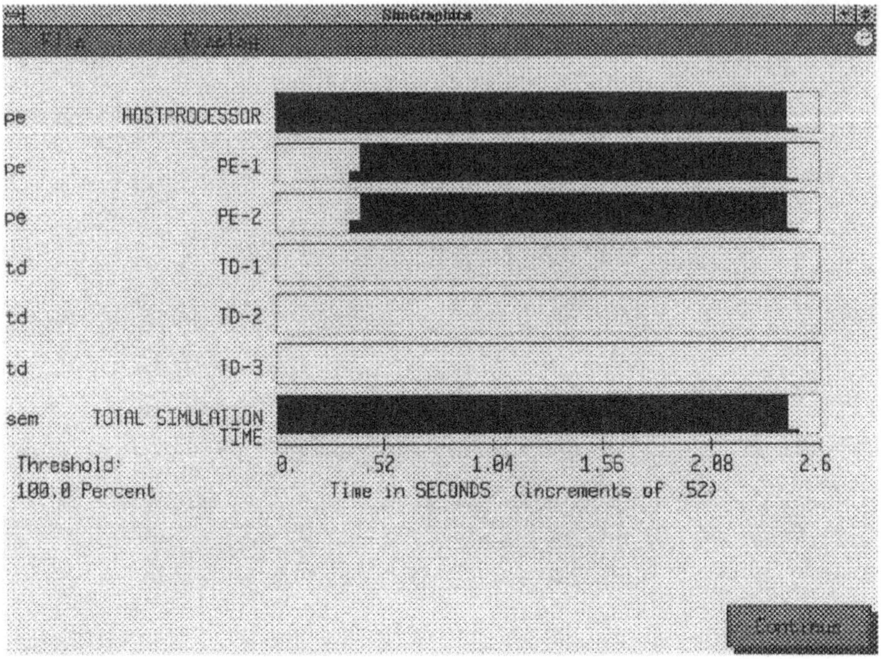

Figure 5.14 Time-line status graph

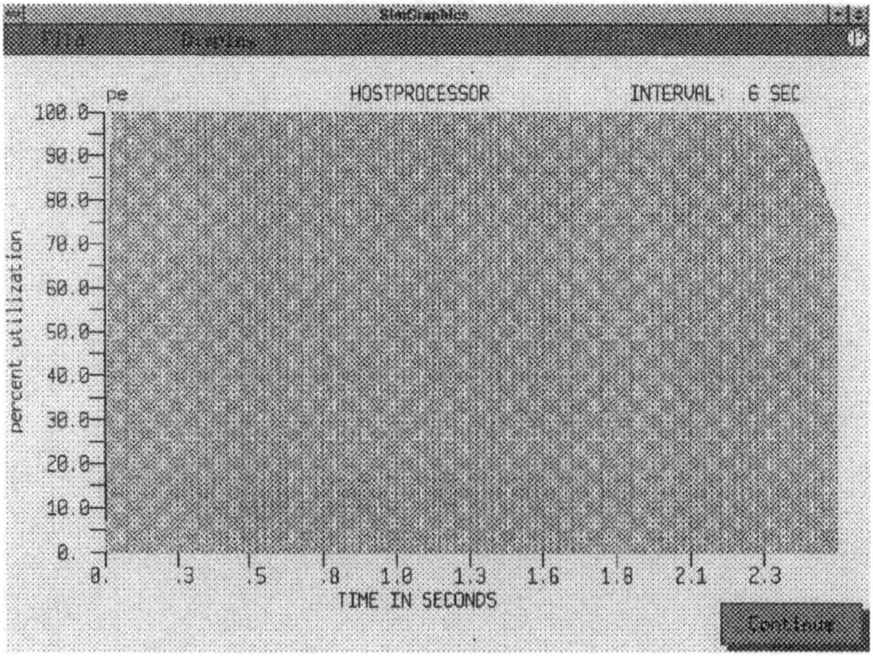

Figure 5.15 a Utilisation graph of host processor

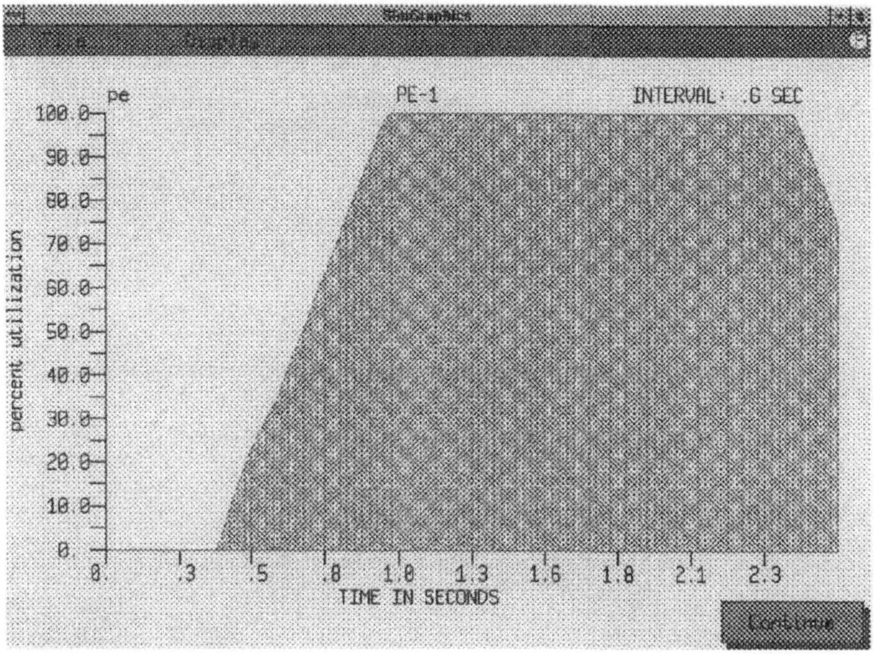

Figure 5.15 b Utilisation graph of PE-1

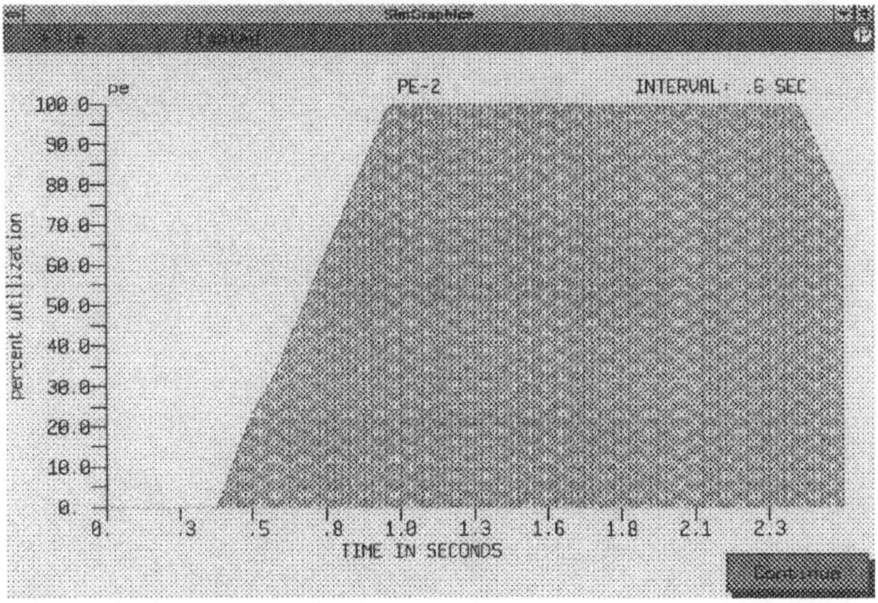

Figure 5.15 c Utilisation graph of PE-2

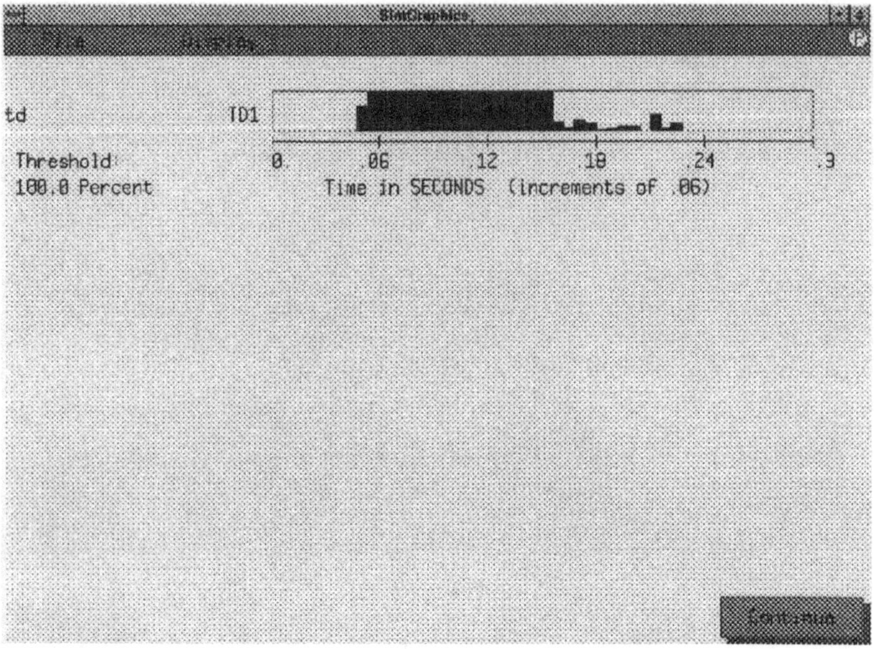

Figure 5.16 Utilisation of TD-1 in Shared Memory Architecture

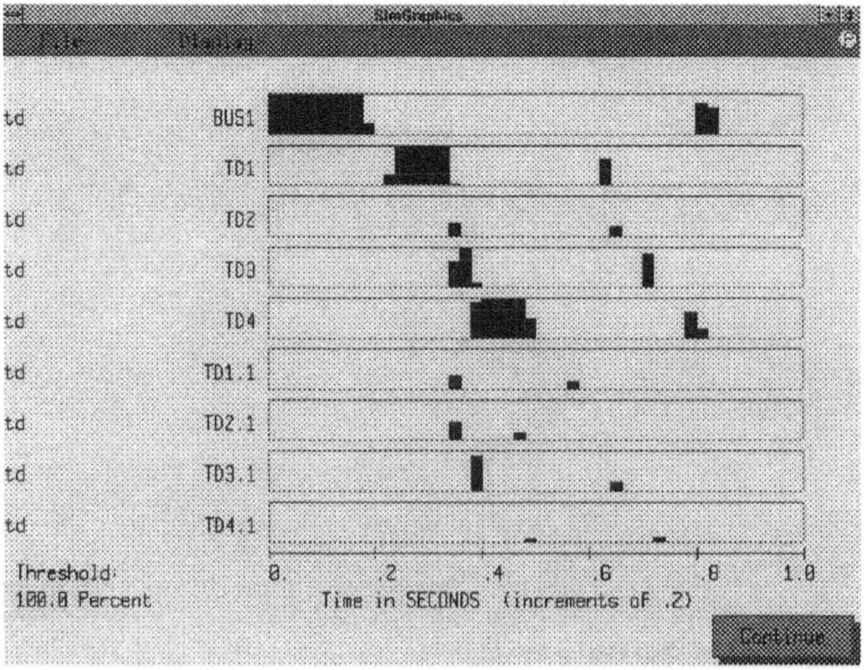

Figure 5.17 Utilisation of transfer devices on Distributed Memory Architecture with Pyramid Topology

6 CONCLUSIONS

6.1 Introduction

This chapter gives the main conclusions derived from the research work presented in this book and some directions for future research.

The DCT-based JPEG algorithm was chosen for this research because it is one of the most widely used image compression algorithms, and because the results could be applied to the MPEG algorithm as well. An enhancement to the JPEG algorithm, called the Block Comparator Technique was introduced. It was shown that the Block Comparator Technique increases the speed of compression operation and improves the compression ratio.

To study the effectiveness of parallel processing, the JPEG algorithm was implemented on three parallel computer architectures, viz., the Mercury system, the Shiva system and the Param system. From the experimental results obtained on these three architectures, it was shown that the system with hybrid memory architecture, gives the best performance in terms of speedup, scaleup and efficiency.

A number of parallel processing Plans were simulated using the Network II.5 simulation package. From the results obtained for different options, it was shown that the execution time obtained for the Non-Block Comparator Technique is the least on the shared memory architecture with global and local memory, and scaleup on the pyramid topology is higher than that for other architectures.

With the widespread application of multimedia systems and increasing data traffic due to the transmission of still and moving pictures, compression of image data has become very important. The main aim of this project was to investigate techniques for improving the performance of the DCT-based JPEG algorithm. There are three main components of this research project, namely:

1) An enhancement to the JPEG algorithm was proposed. This enhancement is called the Block Comparator Technique. The Block Comparator Technique was analysed and the improvements in the performance of the JPEG algorithm were investigated.

2) The JPEG algorithm was implemented on three different parallel machines. Improvement in the performance of the JPEG algorithm on the various architectures was studied.
3) A more detailed study of parallel processing of the JPEG algorithm was carried out by using discrete event simulation.

The different methods involved in the Block Comparator Technique and the conclusions derived from these methods are discussed in section 6.2. The conclusions derived from the experimental results obtained on the three parallel systems are given in section 6.3. A number of important implementation options were simulated using the Network II.5 simulation package. The conclusions derived for the simulation results are discussed in section 6.4. Section 6.5 gives some directions for future research.

6.2 Block Comparator Technique Enhancement to the JPEG Algorithm

Based on the algorithm used, digital image compression techniques can be broadly classified as Vector Quantisation (VQ), Fractal, Wavelet and Discrete Cosine Transform (DCT) techniques. The digital image compression technique developed by the Joint Photographic Experts Group (JPEG) is based on the Discrete Cosine Transform. JPEG technique is widely used in a large variety of applications. Therefore, the JPEG algorithm was chosen for this research purpose.

In the JPEG algorithm all image blocks are processed individually. These blocks of compressed image are stored sequentially. In many types of images, there is the possibility of having one or more similar blocks in the image. Improvement in performance of the compression algorithm can be achieved by locating similar blocks in the image. The Block Comparator Technique was developed to enhance the performance of the JPEG algorithm.

With the help of mathematical analysis it was shown that the Block Comparator Technique improves the speed of compression and reduces the size of the compressed data file. Conclusions derived for the speed of operation are given in section 6.2.1. Conclusions derived for image compression are given in section 6.2.2.

6.2.1 Speed of Operation

There are many methods for implementing the Block Comparator Technique operation. Two options for the block comparison step were selected. The first method involves direct sample-by-sample comparison of all the blocks in an image. The second method consists of sample summation, intensity comparison and sample-by-sample comparison steps. Execution time of the JPEG algorithm was calculated in terms of the number of arithmetic operations such as additions, subtractions, multiplications, divisions, and comparisons. Each arithmetic operation was equated to a number of Base Operations. The number of equivalent Base Operations for each arithmetic operation can be determined for specific processors. Transputer IMS T805 processor was selected for calculating the number of Base Operations.

Improvement in the speed of Block Comparator Technique over the JPEG algorithm (called as Non-Block Comparator - NBCT in this book) was expressed in terms of Speed Improvement Factor (SIF). Conclusions are derived by comparing SIF values obtained for the two methods, and also by comparing SIF values for each method individually. These conclusions are given in the following sections.

Common conclusions for the two methods: The Speed Improvement Factor (SIF) is less than one for zero number of similar blocks (NSB) irrespective of the method used and the image size. This result is expected, as there is no speed improvement when there are no similar blocks; because there is additional computation time required for the block comparison step. Therefore, the Block Comparator Technique (BCT) will add unwanted computational overhead if there are no similar blocks in an image.

SIF values for all image sizes increase monotonically with increase in the value of NSB. For most of the methods studied SIF is greater than one for NSB >= 10%. This indicates that the Block Comparator Technique delivers dividends even for a small number of similar blocks.

Conclusions for the Selection Sort method: For the Selection Sort method SIF values for Number of Blocks (NB) = 256 is greater than one, for NSB in the range of 10% to 100%. Whereas, for NB = 4266 SIF is less than one even upto 50% similar blocks. On the other hand for NB = 15594 SIF is less than one for all values of NSB. This indicates that there is no benefit in using the Block Comparator Technique in conjuction with the Selection Sort method for large images. Selection Sort method is suitable only for small image size.

Conclusion for the Divide and Conquer Sort method: For the Divide and Conquer sort method SIF is greater than one for NSB >= 10% and increases monotonically with NSB. SIF is almost equal for all image sizes for the same values of NSB. Therefore, the Divide and Conquer method is suitable for all image sizes.

Conclusion for the Sample-by-Sample Comparison method: When similar blocks are matched using sample-by-sample comparison, the SIF values for all image sizes and values of NSB are almost equal to the SIF values for the Divide and Conquer method, except for NSB = 100%. For NSB = 100%, SIF is greater for the Sample-by-Sample comparison method than that for the Divide and Conquer Sort method.

Divide and Conquer Sort versus Sample-by-Sample Comparison: Block comparison using Divide and Conquer Sort method is used to sort the image blocks according to the intensities of these blocks. This helps in grouping the blocks of equal intensities. Equal intensity blocks can then be distributed on a parallel computer equally to all the CPUs to balance the work load on all processors. This improves the speedup obtained by parallel processing.

The Block Comparison Technique using Sample-by-Sample comparison method cannot be used for grouping of blocks with equal intensity values. Therefore, we can say that the Block Comparator Technique using Divide and Conquer method is more suitable for parallel processing.

6.2.2 Image Compression Ratio

The Image Compression Ratio (ICR) was calculated for Digital Image Compression Techniques with and without the proposed Block Comparator Technique. For the Block Comparator Technique, the Compressed Image Data Structure can be represented in many formats. Three different structures were selected for the analysis: Compressed Image Data Structure -1 (CIDS-1), CIDS-2 and CIDS-3.

The conclusions derived by comparing ICR values from all Compressed Image Data Structures and by comparing ICR values from each CIDS separately are given in the following sections.

Conclusions for all CIDSs: ICR increases as the image size increases irrespective of the quality of a output image for all CIDSs. For each image size the ICR increases with a decrease in the quality of the output image.

Conclusion for CIDS-1: For the CIDS-1 and quality = 100% ICR is slightly greater than the same for the Non-Block Comparator Technique. For quality = 75% ICR is almost equal to that for the Non-Block Comparator Technique. For quality = 50% and 25% ICR values are less compared to the same for the Non-Block Comparator Technique. This indicates that the size of the compressed image using Compressed Image Data Structure-1 for the Block Comparator Technique is reduced only for image quality greater than 75%.

Conclusions for CIDS-2: By comparing the ICR values of CIDS-2 and CIDS-1, ICR values for CIDS-2 are less than those for CIDS-1 for all image sizes. This is because all Unique Block Numbers are stored in the Unique Block Group. This data structure is more robust than the CIDS-1 data structure, because all the blocks numbers are included in the data structure.

By comparing ICR values of CIDS-2 and the Non-Block Comparator Technique, it was shown that the ICR values of CIDS-2 for quality = 100% are almost equal to that of the Non-Block Comparator Technique. For quality less than 100%, ICR values are less than the same for the Non-Block Comparator Technique.

Conclusions for CIDS-3: By comparing ICR values of CIDS-3 with those for the other two data structures, the ICR values for CIDS-3 are greater in all cases.

By comparing CIDS-3 with the Non-Block Comparator Technique, we can see that the Image Compression Ratio values for CIDS-3 for quality = 100% and 75% are greater than the same for the Non-Block Comparator Technique. For quality = 50% the values are almost equal. This indicates that the CIDS-3 data structure is better than the others for quality greater than 50%.

By comparing all three Compressed Image Data Structures, we can say that CIDS-3 is the best of the three data structures in terms of compression ratio. Therefore, the CIDS-3 data structure was chosen to measure the speed improvement over the Non-Block Comparator Technique. Though CIDS-2 is more robust than either of the other two data structures.

Image Compression Ratio Improvement Factor (ICRIF): The improvement in the compression ratio can be represented by the Image Compression Ratio Improvement Factor (ICRIF).

From the ICRIF graph for CIDS-3 using NSB = 75%, we can see that the ICRIF is almost equal for all image sizes irrespective of the quality of the output image. There is no benefit in using the Block Comparator Technique for images with less than 50% quality. By using the Block Comparator Technique we can get an improvement of 2.8 times over the Non-Block Comparator Technique for quality = 100%.

From these speed and compressed image size comparisons we can say that the Block Comparator Technique is a useful addition to enhance the JPEG compression algorithm. The same Block Comparator Technique can be implemented on parallel computers to speedup the operation even further.

6.3 Implementation of the Digital Image Compression Algorithm

The JPEG digital image compression algorithm was implemented in a variety of ways on three parallel computers. Each uniquely identifiable way of implementation is called a 'Plan' in this book. Each Plan is specified as a 6-tuple consisting of image compression technique, block dependency, image partitioning method, memory architecture, network topology and the number of processors. Some of these Plans were implemented on available parallel computers and other Plans were simulated using the Network II.5 simulation package. Performance of these Plans was evaluated in terms of speedup, scaleup, and efficiency.

Parallel computers can be classified based on memory access technique, network topology and some other issues. Three parallel computers were selected each with a different memory architectures, viz. Mercury system with a distributed memory architecture, Shiva system with a shared memory architecture and Param system with a hybrid memory architecture. Non-Block Comparator Technique was implemented on various number of processors on above three systems and the conclusions derived from the experimental results are given in following section.

6.3.1 Performance Comparison of Digital Image Compression on Three Parallel Computer Architectures

On the Mercury system the Helios Operating system was used as the parallel programming environment. Helios has four levels of communication routines. JPEG algorithm was implemented using two of these, namely, POSIC (Portable Operating Set Instruction Codes) and MPP (Message Passing Primitives). By comparing POSIC and MPP communication routines, we concluded that the speedup for MPP communication routines is higher than that for the POSIC communication routines. But the POSIC communication routines have greater functionality than the MPP communication routines.

The JPEG algorithm was implemented on the Shiva and Param systems each with three processors. The execution times for four processor to seventeen processor systems were estimated by extrapolating the results, obtained on a single processor system, with the help of Gantt charts.

By comparing the results obtained we concluded that the hybrid memory architecture based Param system gave the best results in terms of execution time, speedup, scaleup and efficiency. Nonetheless, the JPEG algorithm is suitable for implementation on distributed as well as shared memory systems.

6.4 Simulation of Digital Image Compression Techniques

Network II.5 simulation package was used for the simulation of digital image compression techniques on parallel architectures.

A total of more than four hundred models were simulated. The execution times obtained from the simulation are tabulated for each model in Appendix A. Each simulated model was examined with respect to execution time, and its derivatives, namely, speedup, scaleup and efficiency. The knowledge gained from these models enabled a comparison of the performance of digital image compression algorithm on parallel computers.

The conclusions derived from the execution times obtained are discussed in section 6.4.1. Performance comparison in terms of speedup, scaleup and efficiency is discussed in section 6.4.2.

6.4.1 Execution Times

The simulation models were first divided into two sets based on the block dependency method, namely, Non-Inter-Processor Communication (NIPC) method and Inter-Processor Communication (IPC) method. The conclusion derived from the NIPC Plans are given in section 6.4.1.1. Conclusions derived from the IPC Plans are given in section 6.4.1.2. A total of twenty models were simulated for the BCT Plans using Number of Similar Block (NSB) = 10%, 30%, 50%, and 75%. The conclusion derived from the execution times obtained for these Plans are given in section 6.4.1.3.

6.4.1.1 Execution Times Obtained for Non-Inter-Processor Communication Plans : By comparing the execution times obtained for the NBCT Plans and the BCT Plans, we can see that the execution time for the BCT Plans is less compared to the execution times for the NBCT Plans upto some Number of Processors (NP), after that the execution time is higher for the BCT Plans. This indicates that the BCT fails to be effective on a large number of processors. The reason for this is explained below.

In the simulation models, the divide and conquer method was used for comparing blocks in an image. In all of the Plans the block comparison step took place on the host processor leading to sequential processing of the block comparison process. Thus, the time taken for the BCT is higher than that for the NBCT as the number of processors increases.

The least execution time was obtained for the NBCT and the BCT Plans on a shared memory architecture with local-plus-global memory organisation.

6.4.1.2 Execution Times for Inter-Processor Communication Plans : By comparing three Inter-Processor Communication Plans it was shown that the execution times obtained on the Distributed Memory Architecture with the Pyramid topology is the lowest.

By comparing the execution times obtained for the NIPC Plans and the IPC Plans on a distributed memory architectures, it can be seen that the IPC method is more efficient for low number of processors whereas the NIPC method becomes more efficient for higher number of processors. Because as the number of processors increases, the communication time taken for transferring neighbouring samples among processors increases for the IPC method. This leads to increase in overall execution time.

6.4.1.3 Execution Times for the BCT with Different NSB values: The Block Comparator Technique was simulated for Number of Similar Blocks = 10%, 30%, 50% and 75%. By comparing execution times obtained for these Plans it was shown that speed improves with an increase in the NSBs. The speed improvement figures obtained for the simulation experiments, for the various NSB values, matched closely to the analytically obtained values.

Highest speed improvement was obtained on twenty one processors connected in a Pyramid topology.

6.4.2 Performance Comparison

Performance of parallel computer architectures was measured in terms of speedup, scaleup and efficiency. Section 6.4.2.1 gives the conclusions derived from speedup comparison. Scaleup comparison is given in section 6.4.2.2 and efficiency comparison is given in section 6.4.2.3.

6.4.2.1 Speedup Comparison:

- On the shared memory architecture higher speedup is obtained with local-plus-global memory organisation for all image sizes.
- The highest speedup is achieved on a distributed memory architecture with the Pyramid topology.
- Higher number of similar blocks in a image lead to higher speedup.

6.4.2.2 Scaleup Comparison

- Scaleup also is the highest on the Pyramid topology. This indicates that both techniques (NBCT and BCT) scaleup well on the pyramid topology.
- The scaleups values for the Block Comparator Technique for all image sizes are lower as compared to the scaleup values for the Non-Block Comparator Technique. Thus, the Block Comparator Technique does not scaleup as well as the Non-Block Comparator Technique.

6.4.2.3 Efficiency Comparison: The highest number of processors at which efficiency is greater than 50% is called as Efficiency Cutoff Point (ECP) in this book. In other words, from cost-benefit analysis point of view a parallel processor with the 'number of processors' = ECP would be the most cost effective.

- The ECP value is the highest on the shared memory architecture with local-plus-global memory organisation for the NIPC method.
- The ECP value for the Pyramid topology is the highest for the IPC method.

6.5 Directions for Future Research

When this project was initiated a few years ago, low cost hardware implementations of the JPEG and the MPEG standards were not readily available. At the present time JPEG and MPEG chips/cards are readily available at reasonably low cost. But these hardware devices are fixed for a specific standard. With rapid advancement in compression technology a flexible image compression scheme is required. In such a flexible scheme the two ends can negotiate the standard, and the parameters to be used for image compression. For this scheme to work image compression and decompression must be performed in software. To be able to perform real-time motion picture compression and decompression in software parallel processing can be employed.

The research work carried out in this project can be extended to include motion picture compression. Some more specific directions for future work are presented below.

1. Speed of Operation: The Selection Sort and the Divide and Conquer methods were used for the block comparison step in the Block Comparator Technique. By comparing the Speed Improvement Factors it was found that the Divide and Conquer method is better than the Selection Sort method. But, there may be other sort methods that perform better than the Divide and Conquer sort method for this application. Therefore, future research can explore other sorting methods which may be faster than the Divide and Conquer sort method, especially for parallel processing.

2. Quality of Service: Three Compressed Image Data Structure were used for the Block Comparator Technique. From these Image Data Structures, the CIDS-2 was found to be most robust, though the compressed image is slightly larger as compared to one of the other structures (CIDS-3). This robustness is desirable for providing good quality of service in many applications such as video-on-demand. Future research can explore ways of reducing the compressed image size of CIDS-2.

In the CIDS-2 all blocks include the block numbers. This helps to identify lost blocks. If any of the blocks is missing, this block is replaced by an empty block. This leads to blockiness in the image. Future research can focus on recovery of missing blocks during decompression operation to reduce this blockiness effect, so that the decompressed image can be of better quality.

3. Reliability : In this research all processors in the parallel system were considered to be operational. But there is a chance of breakdown of one or more of the processors during task allocation, compression and collection of compressed data. Future research can include development of reliable parallel processing techniques specifically for image compression.

4. Parallel Block Comparison: In our study the block comparison step took place on the host processor leading to sequential processing of the same. This leads to the fact that the time taken for the Block Comparator Technique is higher on a large number of processors as compared to the Non-Block Comparator Technique. Execution time can be reduced further by parallelising the block comparison step.

REFERENCES

Anderson, 90. Anderson, M. S. ; Drewer, P. C., "Design Overview of the Shiva", Proceedings of the IEEE TENCON'90: 1990 IEEE Region 10 Conference on Computer and Communication Systems, Hong Kong, 24 - 27 Sept. 1990, Vol. 1, pp. 155 - 159.

Anderson, 92. Anderson, m. ; Yesberg, J. D. ; Yakovleff, A. J. et. al., "A Heterogeneous Parallel Accelerator for Image Analysis and Radar Signal Processing", Proceedings of the twenty fiveth Hawaii International Conference on System Sciences, Kauai, HI, USA, 7 - 10 Jan. 1992, Vol. 1, pp. 129 - 138.

Ang, 91. Ang, P. H. ; Ruetz, P. A. ; Auld D., "Video Compression Makes Big Gains", IEEE Spectrum, Oct. 1991, pp. 16 - 19.

Aravind, 89. Aravind, R. ; Cash, G. L. ; Worth, J. P., "On Implementing The JPEG Still-Picture Compression Algorithm", SPIE Visual Communications and Image Processing IV, Vol. 1199, 1989, pp. 799 - 808.

Baran, 90. Baran, N., "Putting the Squeeze on Graphics", Byte, Dec. 1990, pp. 289 - 294.

Barnsley, 93. Barnsley, M. F. ; Hurd, L. P., "*Fractal Image Compression*", A. K. Peters Ltd., Wellsley, 1993.

Bevinakoppa, 92. Bevinakoppa, S. G. ; Sharda, H. N. ; Hulskamp, J. ; Sharda, N. K., "Digital Image Compression on a Network of Transputers", Transputers and Parallel Applications Conference, Melbourne 4-5 Nov. 1992, pp 25-32.

Bevinakoppa, 94a. Bevinakoppa, S. G. ; Sharda, N. K. ; Sharda, H. N., "Performance analysis of a parallel JPEG algorithm". PART'94: Australian Workshop on Parallel and Real Time system, 7-8 July 1994, Melbourne, pp. 42 - 52.

Bevinakoppa, 94b. Bevinakoppa, S. G. ; Sharda, N. K. ; Sharda, H. N., "Implementation of JPEG Algorithm on Shiva Parallel Architecture", IWPP: First International Workshop on Parallel Processing, Bangalore, India, Dec. 27-31 1994, pp. 184 - 188.

Bevinakoppa, 95. Bevinakoppa, S. G. ; Sharda, N. K. ; Sharda, H. N., "Parallel Implementation of DCT-based Digital Image Compression on a Param System", PART'95: Australasian Workshop on Parallel and Real Time system, 7-8 August 1995, Perth, pp. 401 - 407.

Bhatkar, 91. Bhatkar, V. P., "*Advanced Computing: Proceedings of the Centre for Development of Advanced Computing*", Eds. Bhatkar, V. P., Joshi, A. V., Sharma, A. K., Tata Mc-Graw Hill Publishing Corporation Ltd., New Delhi, India, August 1988 - July 1991.

Bratley, 83. Bratley, P. ; Fox, B. L. ; Schrage, L. E., "*A Guide to Simulation*", Springer-Verlag, Newyork, 1983.

Browne, 89. Browne, R. F. ; Hodgson, R. M., " Mapping Image Processing Operations onto Transputer Networks", 1989, Microprocessor and Microsystems.

Burns, 89. Burns, A. ; Wellings, A., "*Real-time systems and their programming languages*", Addison-Welsy publishing company, 1989.

CACI, 94. CACI Products Company, "*Network II.5 User's Manual*", 1994.

Chong, 90. Chong, M. N. ; Soraghan, J. J., " Transputer Based Quadtree Data Structure for Adaptive Transform Coding". Signal Processing V: Theroies and Applications, 1990, pp. 1563-1566.

Chung-Ta King, 91. Chung-Ta King, "Skewed Partition - Theory and Practice", 1991 IEEE Conference.

Cockroft, 91. Cockroft, G. ; Hourwitz, L., "NEXTstep: Putting JPEG to Multiple Uses", Communication of the ACM, April 1991, Vol. 34, No. 4, pp. 45 and 116.

Cornell, 93. Cornell, E. ; Kurz, P., "Wavelets and Improved JPEG Fuel New Generation of Digital Video Engines, AV Video 1993, pp. 40 - 46.

Cosman, 96. Cosman, P. C. ; Gray, R. M. ; and Vetterli, M., "Vector Quantisation of Image Subbands: A survey", IEEE Transactions on Image Processing, Feb. 1996, Vol. 5, No. 2, pp. 202 - 225.

Draft, 90. "*Coding of Moving Pictures and Associated Audio*", Committee Draft Standard - ISO 11172, ISO/MPEG 90/176 Dec. 1990.

Eknath, 91. Eknath, P. R. ; Bhasin, L. ; Degnekar, A. etl., "*Param Parallel Computer*", Advanced Computing: Proceedings of the Centre for Development of Advanced Computing, Eds. Bhatkar, V. P., Joshi, A. V., Sharma, A. K., Tata Mc-Graw Hill Publishing Corporation Ltd., New Delhi, India, August 1988 - July 1991, pp. 71 - 85.

Elliott, 89. Elliott, J. ; Beaumont, J. M. ; Grant, P. M. et. al., "Real Time Videophone Image Algorithm on a Concurrent Supercomputer". British Telecom Research Laboratories.

Fleming, 88. Fleming, P. I., "*Parallel Processing in Control: the Transputer and Other Architectures*",Peter Peregrinus Ltd., London, United Kingdom, 1988.

Furht, 94. Furht, B., "Multimedia Systems: An Overview", IEEE Multimedia, Vol.1, No.1, Spring 1994, pp. 47 - 59.

Furht, 95a. Furht, B., "*A guided Tour of Multimedia Systems and Applications*", IEEE Computer Society Press Lab, Alamitos, California, 1995.

Furht, 95b. Furht, B. " *A Survey of Multimedia Techniques and Standards- JPEG Compression*", Journal of Real-Time Imaging, Vol. 1, No. 1, April 1995.

Gall, 91. Gall, D., "MPEG: A Video Compression Standard for Multimedia Applications", Communications of the ACM, April 1991, Vol. 34, No. 4, pp. 47-58.

Geetha, 91. Geetha, S. ; Kumar, P. ; Sandya, V., "Parallel Programming with Paras", C-DAC Report, pp. 1 - 19.

Gersho, 92. Gersho, A. ; Gray, R. M., "*Vector Quantisation and Signal Compression*", Kluwer Academic Publishers, Boston, 1992.

Harney, 91. Harney, K. ; Keith, M. ; Lavelle, G. et. a;., "The i750 Video Processor: A Total Multimedia Solution", Communications of ACM, Vol. 34, No.4, April 1991, pp. 65 - 79.

Hemery, 91. Hemery, F. ; Lazure, D. ; Delattre, E. et. al., " An Analysis of Communication and Multiprogramming in the Helios Operating System", Microprocessing and Microprogramming, 32, 1991, 137-144.

Hord, 93. Hord, R. M., "*Parallel Supercomputing in MIMD architectures*", CRC press, Florida, 1993.

Hunt, 93. Hunt, J. C. ; Kevlahan, N. K. ; Vassilicos, J. C. et. al., "Wavelets, Fractals, and Fourier Transforms", Ed. Farge, M. ; Hunt, J. C. ; Vassilicos, J. C., Clarendon Press, Oxford, 1993, pp. 1- 37.

Ian, 92. Ian D., "*Helios operating system*", Perihelion Software Limited, 1992.

INMOS, 89. INMOS ltd. "*INMOS: Transputer Databook*", 1989.

Intel, 90. i860 *Hardware Reference Manual*, Intel, 1990.

Jain, 89. Jain, A. K., "*Fundamentals of digital image processing*", Prentice Hall, 1989.

Kajiwara, 92. Kajiwara, K., "JPEG compression for PACS", Computer Methods and Programs in Biomedicine, No. 37, 1992, pp. 343-351.

Karnak, 92a. Karnak, D. A. et. al., "Shiva Mark I - Detailed Hardware Design", ITD Divisional Paper, Adelaide, 1992.

Karnak, 92b. Karnak, D. A. ; Yakovleff, A. J. ; Yesberg, J. D. et. al., "Shiva Mark II Hardware Architecture, Virsion 1", ITD Divisional Paper, Adelaide, May 1992.

Kinoshita, 92. Kinoshita, T ; Nakahashi, T., "A 130 Mb/s Compact HDTV CODEC Based on a Motion-Adaptive DCT Algorithm", IEEE Journal on Selected Areas in Communications, Vol. 10, No. 1, Jan. 1992, pp.122 - 129.

Koornwinder, 93. Koornwinder, T. H., "*Wavelets: An Elementary Treatment of Theory and Applications*", World Scientific Publishing Co. Pte. Ltd., 1993.

Krishnamurthy, 89. Krishnamurthy, E. V., "Parallel Processing Principles and Practice", Addison-Wesley Publishing Company, Singapore, 1989.

Kruse, 94. Kruse, R. L., "*Data Structures and Program Design*", Prentice Hall, Englewood Cliffs, New Jersey, 1994.

Kumar, 92. Kumar, M. K. ; Kumar, P. S. ; Basu, A., "A Library Environment for Distributed Memory Multiprocessors", Proceedings. Sixth International Parallel Processing, Beverly Hills, CA, USA, 23 - 26 March 1992, pp. 483 - 486.

Kumar, 94. Kumar, V. et al., "*Introduction to Parallel Computing: Design and Analysis of Algorithms",* Redwood City, California, Benjamin/Cumming, Publishing Company, 1994.

Leger, 91. Leger, A. ; Omachi, T. ; Wallace, G. K., "JPEG Still Picture Compression Algorithm", Optical Engineering, July 1991, Vol. 30, No. 7, pp.947 - 954.

Leigh, 83. Leigh, J. R., "*Modelling and Simulation*", Peter Peregrinus Ltd., London, UK, 1983.

Leonard, 91. Leonard, M., "IC Executes Still-Picture Compression Algorithms", Electronic Design, May 23, 1991, pp. 49 - 53.

Lewis, 92. Lewis, T. G. ; El-Rewini, H., "*Introduction to Parallel Computing*", Prentice Hall, 1992.

Liou, 91. Liou, M., "Overview of the Px64 Kbits/s Video Coding Standard", Comm. of the ACM, Vol. 34, No. 4, April 1991, pp. 59-63.

Margulis,90. Margulis, N., *i860 Microprocessor Architecture*, Osborne, 1990.

Maurer,88. Maurer, P. M., "The Dataflow Model of Computation in an Enhanced Von Neumann Processor", IEEE Symposium on Parallel Processing, 1988, pp. 235 - 239.

Mitchell, 90. Mitchell, D. A. ; Thompson, J. A. ; Manson, G. A. et. al., "*Inside The Transputer",* Backwell Scientific Publications, 1990.

Murthy, 91. Murthy, T. S. ; Eknath, P. R., "Param Vector Facility", Advanced Computing: Proceedings of the Centre for Development of Advanced Computing, Eds. Bhatkar, V. P., Joshi, A. V., Sharma, A. K., Tata Mc-Graw Hill Publishing Corporation Ltd., New Delhi, India, August 1988 - July 1991, pp. 86 - 89.

Nacken, 93. Nacken, P., "*Image Compression Using Wavelets*", Ed. Koornwinder, T. H., in Wavelets: An Elementary Treatment of Theory and Applications, World Scientific, Singapore, 1993.

Naylor, 68. Naylor, T. H. ; Balintfy, J. L. ; Burdick, D. S. et. al., "*Computer Simulation Techniques*", John Wiley & Sons, Inc., New York, 1968.

Nelson, 92a. Nelson, M., "*The Data Compression Book: featuring fast, efficient data compression techniques in C* ", M&T Books, 1992.

Nelson, 92b. Nelson, M. ; Cavaiuolo, M. ; Yakovleff, A., "An Architecture for Real-Time 3D Graphical Simulation", International Conference on Automation, Robotics and Computer Vision, Singapore, Sept. 1992, pp. CV.17.5.1 - CV.17.5.5.

Nelson, 93a. Nelson M. ; Yakovleff, A., "Shiva Programming Notes", Defence Science Technology and Organisation Technical report, Adelaide, 27 Sept. 1993.

Nelson,93b. Nelson, M. ; Cavaiuolo, M. ; Yakovleff, A., "A Heterogeneous Architecture for Stereoscopic Visualisation", First IEEE Virtual Reality Anual International Symposium Seattle, USA, SEP. 1993, pp. 349 - 355.

Ogawa, 92. Ogawa, K. ; Urano, T. ; Konda, K. et al., "A single chip Compression/Decompression LSI Based on JPEG", IEEE Transactions on Consumer Electronics, Vol. 38, No. 3, Aug. 1992, pp. 703- 710.

Papathanassiadis, 92. Papathanassiadis T. “Image block partitioning: A Compression Technique suitable for Parallel Processing”, Conference publication, Nov. 1992.

Pennabaker, 93. Pennabaker, W. B. ; Mitchell, J. L., "*JPEG Image Data Compression Standard*", Van Nostrand Reinhold, New York, 1993.

Pountain, 87. Pountain, D., "A *Tutorial Introduction to OCCAM Programming*", Inmos Ltd., 1987.

Quinnell, 93. Quinnell, R. A., "Image Compression Part2", EDN Design Feature, March 4, 1993, pp. 120 - 126.

Ram, 91. Ram, N. M. ; Perianayagam, K. S. ; Morra, R. etl., "A High Performance Parallel System Architecture", Advanced Computing: Proceedings of the Centre for Development of Advanced Computing, Eds. Bhatkar, V. P., Joshi, A. V., Sharma, A. K., Tata Mc-Graw Hill Publishing Corporation Ltd., New Delhi, India, August 1988 - July 1991, pp. 663 - 675.

Ramaswamy, 93. Ramaswamy, S. V. ; Miller, G. D., “Multiprocessor DSP Architecture That implement The FCT Based JPEG Still Picture Image Compression Algorithm With Arithmetic Coding”, IEE Transactions on Consumer Electronics, Vol.39, No.1, Feb. 1993, pp. 1- 5.

Rao, 91. Rao, C. M. ; Ram, M. N. ; Perianayagam, K. S. etl., "MTK/860 - A Multi Threading Kernel for the i860", Advanced Computing: Proceedings of the Centre for Development of Advanced Computing, Eds. Bhatkar, V. P., Joshi, A. V.,

Sharma, A. K., Tata Mc-Graw Hill Publishing Corporation Ltd., New Delhi, India, August 1988 - July 1991, pp. 676 - 682.

Rashinkar, 91. Rashinkar, P. ; Bhasin, L. ; Balachandran, S. etl., "Parallel Processing Application Accelerators", Advanced Computing: Proceedings of the Centre for Development of Advanced Computing, Eds. Bhatkar, V. P., Joshi, A. V., Sharma, A. K., Tata Mc-Graw Hill Publishing Corporation Ltd., New Delhi, India, August 1988 - July 1991, pp. 90 - 95.

Rinaldo, 95. Rinaldo, R, ; Calvagno, G., "Image Coding by Block Prediction of Multiresolution Subimages", IEEE Transactions on Image Processing, Vol. 4, No. 7, July 1995, pp. 909 - 920.

Reitman, 81. Reitman, J., "*Computer Simulation Applications*", Robert E. Krieger Publishing Company, Florida, 1981.

Roberts, 83. Roberts, N. ; Anderson, D. ; Deal, R. et. al., "*Introduction to Computer Simulation: A System Dynamics Modelling Approach*", Addison-Wesley Publishing Company, 1983.

Ruetz, 93. Ruetz, P. A. ; Tong, P. ; Luthi, D. A. et al., " A Video Rate JPEG Chip Set", Journal of VLSI Signal Processing, Vol. 5, 1993, pp. 141 - 150.

SGS_Thomson, 91. SGS_Thomason, *" The T9000 Transputer Products Overview Manual",* Inmos Ltd. 1991.

Sharda, 93. Sharda, N. K. ; Bevinakoppa, S. G. ; Sharda, H. N., "Parallel Implementation of Digital Image Compression Based on the JPEG Standard", Technical report 32 COMP 6, Department of Computer and Mathematical Sciences, Victoria University of Technology, Nov. 1993.

Sharda, 95. Sharda, N. K., "*Information Networking*", ACCT press, 1995.

Siegel, 85. Seigel, H. J., "*Interconnection Networks for Large-Scale Parallel Processing",* McGraw-Hill Publishing Company, USA, 1985.

Sijsterman, 91. Sijstermans, F. ; Vander Meer, J., "CD-I Full-Motion Video Encoding on a Parallel Computer", Communications of the ACM, Vol.34, No. 4, April 1991, pp. 81 - 91.

Srivastava,91. Srivastava, A. K. ; Kshetramade, S. C., PRESHAK: a Generic Tool to Implement Application Specific Message-Passing Communication Kernels for Concurrent Machines, Proceedings. The fifth International Parallel Processing, Symposium, Anaheim, CA, USA, 30 April - 2 May 1991, pp. 626 - 629.

Stephenson, 71. Stephenson, R. E., "*Computer Simulation for Engineers*", Harcourt Brace Jovanwich, USA, 1971.

Sugai, 87. Sugai, M. ; Kanuma, A. ; Suzuki, K. et. al., "VLSI Processor for Image Processing", Proceedings of the IEEE, Vol, 75, No. 9, Sep. 1989, pp. 1160 - 1165.

Sun, 90. Sun Microsystems, SBus Specification, 1990.

Tinker, 89. Tinker, M, "DVI Parallel Image Compression", Communications of the ACM, Vol. 32, No.7, July 1989, pp. 844 -851.

Tulshibagwale, 94. Tulshibagwale, A. ; Parikh, S. ; Mahajan, S. etl., "The RISAM Storage Manager for Parallel Architectures", Proceedings of the Third International Conference on Parallel and Distributed Information Systems, Austin, TX, USA, 28 - 30 Sept. 1994, pp. 69 - 70.

Udpikar, 91. Udpikar, V. ; Singh, R. K. ; Madhasudan, b. N. etl., "ImagePRO: Transputer Based Interactive Image Procesing Package", Advanced Computing: Proceedings of the Centre for Development of Advanced Computing, Eds.

Bhatkar, V. P., Joshi, A. V., Sharma, A. K., Tata Mc-Graw Hill Publishing Corporation Ltd., New Delhi, India, August 1988 - July 1991, pp. 437 - 443.

Ungere, 91. Ungere, T., " Parallelising C++ Programs for Transputer Systems", 32, 1991, Microprocessing and Microprogramming, pp. 463-70.

Vaaben, 91. Vaaben, J. ; Niss, B., "Compressing Images With JPEG", Information Display 7 and 8, 1991, pp. 12 - 13 and 59.

Wallace, 92. Wallace, G. K., "The JPEG Still Picture Compression Standard", IEEE Transaction on Consumer Electronics, Vol. 38, No. 1, Feb. 1992, pp. xviii - xxviv.

Yakovleff, 91. Yakovleff, A. ; Yesberg, J. ; Karnak, D. et. al., "An Expandable Supercomputer Architecture with Dynamic Reconfigurability Properties", Fourth Australian Supercomputer Conference, Gold Cost, Dec. 1991, pp. 175 - 83.

Yakovleff, 94. Yakovleff, A. ; Cavaiuolo, M., "A Simulation Environment for Very Large Neural Networks", IEEE International Conference on Aucoustics, Speech, and Signal Processing, Adelaide, Australia, April 1994, Vol. 2, pp. 577 - 580.

Zomaya, 96. Zomaya, A. Y., "*Parallel Computing: Paradigms and Applications*", International Thomson Computer Press, London, 1996.

APPENDIX A

Table A.1 Execution times for NIPC Plan P2 (NBCT on a Shared Memory Architecture with Global Memory)

Number of Processors-NP	Execution times in msec.		
	For 125 x 125 image	For 625 x 429 image	For 1100 x 900 image
1	112.32	1871.81	6844.12
3	47.15	768.68	3315.63
5	33.29	525.58	1931.63
7	25.84	409.43	1574.19
11	21.12	328.39	1213.44
16	**17.82**	267.73	1018.32
20	18.00	251.81	929.77
25		**235.49**	**869.13**
27		235.72	869.43

Table A.2 Execution times for NIPC Plan P3 (NBCT on a Shared Memory Architecture with Local-plus-Global Memory)

Number of Processors-NP	Execution times in msec.		
	For 125 x 125 image	For 625 x 429 image	For 1100 x 900 image
1	112.32	1871.81	6844.12
3	41.40	699.07	2557.27
5	27.69	476.79	1748.24
7	21.85	380.42	1397.12
11	15.12	278.39	1026.73
16	10.93	226.78	870.31
20	**10.47**	**194.96**	**732.21**
25	10.67	195.44	734.62

Table A.3 Execution times for NIPC Plan P4 (NBCT on a Distributed Memory Architecture with Tree Topology)

Number of Processors -NP	Execution times in msec.		
	For 125 x 125 image	For 625 x 429 image	For 1100 x 900 image
1	112.32	1871.81	6844.12
3	41.40	719.30	2631.21
5	32.82	529.36	1941.19
9	24.96	410.74	1503.83
15	18.77	315.96	1155.50
21	16.63	280.89	1034.26
27	**15.49**	**267.72**	**981.38**
33	16.09	279.42	993.62

Table A.4 Execution times for NIPC Plan P5 (NBCT on a Distributed Memory Architecture with Torus Topology)

Number of Processors-NP	Execution times in msec.		
	For 125 x 125 image	For 625 x 429 image	For 1100 x 900 image
1	112.32	1871.81	6844.12
3	41.40	719.30	2631.21
5	32.82	529.36	1941.19
9	20.75	342.29	1234.28
13	19.60	323.03	1180.84
17	16.16	271.72	995.82
21	**11.69**	**262.58**	**959.29**
26	16.09	266.42	976.67

Table A.5 Execution times for NIPC Plan P6 (NBCT on a Distributed Memory Architecture with Pyramid Topology)

Number of Processors -NP	Execution times in msec.		
	For 125 x 125 image	For 625 x 429 image	For 1100 x 900 image
1	112.32	1871.81	6844.12
5	23.36	386.07	1412.65
9	21.85	359.84	1317.72
21	18.15	309.03	1128.46
37	**13.43**	**224.55**	**812.04**
53	14.61	239.86	876.50

Table A.6 Execution times for NIPC Plan P7 (NBCT on a Distributed Memory Architecture with Cube Topology)

Number of Processors-NP	Execution times in msec.		
	For 125 x 125 image	For 625 x 429 image	For 1100 x 900 image
1	112.32	1871.81	6844.12
5	23.36	386.07	1412.65
9	21.85	359.84	1317.72
28	**16.30**	**277.49**	**959.14**
49	29.38	387.65	1398.27

Table A.7 Execution times for NIPC Plan P8 (NSB = 10%) (BCT on a Shared Memory Architecture with Global Memory)

Number of Processors -NP	Execution times in msec.		
	For 125 x 125 image	For 625 x 429 image	For 1100 x 900 image
1	104.65	1745.62	6382.80
3	47.62	717.63	2625.94
5	30.01	500.03	1830.35
7	24.41	414.76	1483.08
11	18.97	322.38	1179.25
16	**16.62**	**275.64**	**1005.15**
20	18.22	303.42	1109.62

Table A.8 Execution times for NIPC Plan P9 (NSB = 10%) (BCT on a Shared Memory Architecture with Local-plus-Global Memory)

Number of Processors -NP	Execution times in msec.		
	For 125 x 125 image	For 625 x 429 image	For 1100 x 900 image
1	104.65	1745.62	6382.80
3	39.30	659.82	2414.62
5	26.43	440.38	1631.05
7	20.82	351.72	1264.48
11	15.28	260.90	954.54
16	12.76	229.10	800.43
20	**10.73**	**200.42**	**733.24**
25	10.75	217.98	818.73

Table A.9Execution times for NIPC Plan P1 (NSB = 10%) (BCT on a Distributed Memory Architecture with Tree Topology)

Number of Processors -NP	Execution times in msec.		
	For 125 x 125 image	For 625 x 429 image	For 1100 x 900 image
1	104.65	1745.62	6382.80
3	41.07	689.33	2522.44
5	25.41	449.13	1644.43
9	19.93	329.57	1310.62
15	**15.96**	**259.58**	**1030.67**
21	16.84	281.22	1109.22

Table A.10 Execution times for NIPC Plan P10 (NSB = 10%) (BCT on a Distributed Memory Architecture with Torus Topology)

Number of Processors -NP	Execution times in msec.		
	For 125 x 125 image	For 625 x 429 image	For 1100 x 900 image
1	104.65	1745.62	6382.80
3	41.07	689.33	2522.44
5	25.41	449.13	1644.43
9	**21.33**	**348.42**	1373.29
13	23.80	400.83	**1173.70**
17			1325.98

Table A.11 Execution times for NIPC Plan P11 (NSB = 10%) (BCT on a Distributed Memory Architecture with Pyramid Topology)

Number of Processors -NP	Execution times in msec.		
	For 125 x 125 image	For 625 x 429 image	For 1100 x 900 image
1	104.65	1745.62	6382.80
5	22.28	373.23	1366.00
9	19.26	318.53	1165.69
21	**13.59**	**226.41**	**833.01**
37	13.96	239.29	876.25

Table A.12 Execution times for NIPC Plan P12 (NSB = 10%) (BCT on a Distributed Memory Architecture with Cube Topology)

Number of Processors -NP	Execution times in msec.		
	For 125 x 125 image	For 625 x 429 image	For 1100 x 900 image
1	104.65	1745.62	6382.80
5	22.28	373.23	1366.00
9	**19.26**	**318.53**	**1165.69**
28	19.94	334.80	1186.99

Table A.13 Execution times for IPC Plan P13 (NSB = 10%) (BCT on a Shared Memory Architecture with Global Memory)

Number of Processors -NP	Execution times in msec.		
	For 125 x 125 image	For 625 x 429 image	For 1100 x 900 image
1	104.65	1745.62	6382.8
5	42.54	449.90	1653.58
7	25.96	363.67	1294.69
9	22.94	329.36	1164.74
13	**20.09**	**299.42**	**1092.95**
17	21.62	310.06	1125.71

Table A.14 Execution times for IPC Plan P14 (NSB = 10%) (BCT on a Distributed Memory Architecture with Torus Topology)

Number of Processors -NP	Execution times in msec.		
	For 125 x 125 image	For 625 x 429 image	For 1100 x 900 image
1	104.65	1745.62	6382.8
3	42.36	695.46	2542.94
5	30.33	439.70	1628.26
9	20.81	328.12	1143.87
13	**18.80**	**295.36**	**1069.14**
17	19.49	308.95	1123.73

Table A.15 Execution times for IPC Plan P15 (NSB = 10%) (BCT on a Distributed Memory Architecture with Pyramid Topology)

Number of Processors -NP	Execution times in msec.		
	For 125 x 125 image	For 625 x 429 image	For 1100 x 900 image
1	104.65	1745.62	6382.80
5	28.39	453.26	1648.53
9	20.15	324.59	1062.56
21	**12.03**	**220.21**	**830.27**
37	12.47	226.38	865.21

Table A.16 SIF values for NBCT Plan P3 and BCT Plan P9(on a Shared Memory Architecture with Local-plus-Global Memory)

Number of Processors -NP	Speed Improvement Factor - SIF		
	For 125 x 125 image	For 625 x 429 image	For 1100 x 900 image
1	1.07	1.07	1.07
3	1.05	1.06	1.06
5	1.05	1.08	1.07
7	**1.05**	1.08	1.10
11	0.99	**1.07**	1.08
16	0.98	0.99	**1.09**
20	0.98	0.97	1.00
25	0.91	0.90	0.90

Table A.17 SIF values for NBCT Plan P4 and BCT Plan P1(on a Distributed Memory Architecture with Tree Topology)

Number of Processors-NP	Speed Improvement Factor - SIF		
	For 125 x 125 image	For 625 x 429 image	For 1100 x 900 image
1	1.07	1.07	1.07
3	1.01	1.04	1.04
5	1.29	1.18	1.18
9	1.25	1.25	1.15
15	**1.18**	**1.22**	**1.12**
21	0.99	1.00	0.93

Table A.18 SIF values for NBCT Plan P5 and BCT Plan P10(on a Distributed Memory Architecture with Torus Topology)

Number of Processors -NP	**Speed Improvement Factor - SIF**		
	For 125 x 125 image	**For 625 x 429 image**	**For 1100 x 900 image**
1	1.07	1.07	1.07
3	1.01	1.04	1.04
5	**1.29**	**1.18**	1.18
9	0.97	0.98	1.09
13	0.82	0.81	**1.01**
17			0.75

Table A.19 SIF values for NBCT Plan P6 and BCT Plan P11 (on a Distributed Memory Architecture with Pyramid Topology)

Number of Processors -NP	**Speed Improvement Factor - SIF**		
	For 125 x 125 image	**For 625 x 429 image**	**For 1100 x 900 image**
1	1.07	1.07	1.07
5	1.05	1.03	1.03
9	1.13	1.13	1.13
21	**1.34**	**1.36**	**1.35**
37	0.96	0.94	0.93

Table A.20 SIF values for NBCT Plan P7 and BCT Plan P12 **(on a Distributed Memory Architecture with Cube Topology)**

Number of Processors -NP	**Speed Improvement Factor - SIF**		
	For 125 x 125 image	**For 625 x 429 image**	**For 1100 x 900 image**
1	1.07	1.07	1.07
5	1.05	1.03	1.03
9	**1.13**	**1.13**	**1.13**
28	0.82	0.83	0.81

Table A.21 Speedup for NIPC Plan P2(NBCT on a Shared Memory Architecture with Global Memory)

Number of Processors-NP	Speedup-S		
	For 125 x 125 image	For 625 x 429 image	For 1100 x 900 image
1	1	1	1
3	2.38	2.44	2.06
5	3.37	3.56	3.54
7	4.35	4.57	4.35
11	5.32	5.70	5.64
16	**6.30**	6.99	6.72
20	6.24	7.43	7.36
25		**7.95**	**7.87**
27		7.94	7.86

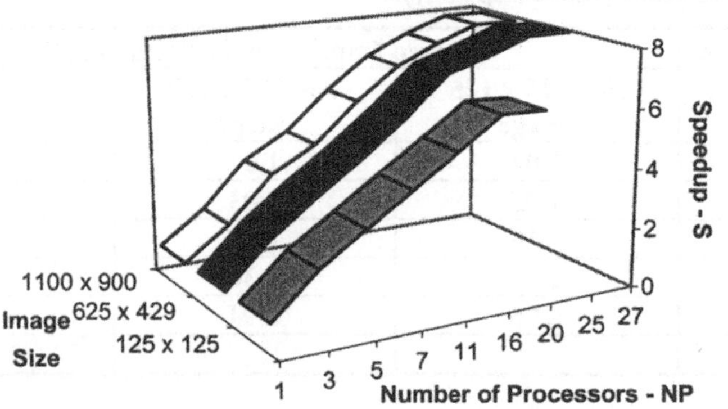

Figure A.1a Speedup graph for Plan P2

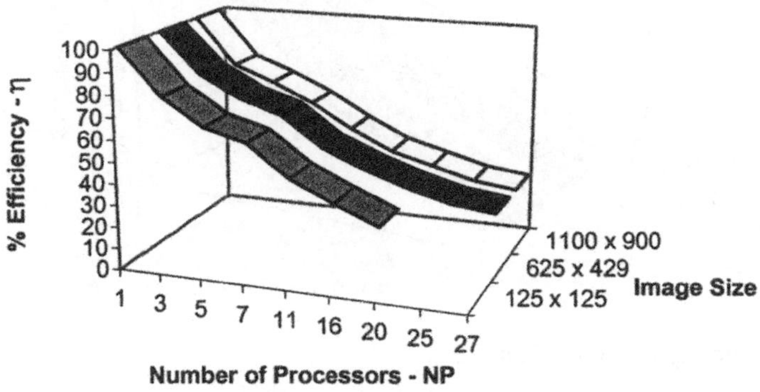

Figure A.1b Efficiency graph for Plan P2

Table A.22Speedup for NIPC Plan P3(NBCT on a Shared Memory Architecture with Local-plus-Global Memory)

Number of Processors -NP	Speedup-S		
	For 125 x 125 image	For 625 x 429 image	For 1100 x 900 image
1	1	1	1
3	2.71	2.68	2.68
5	4.06	3.93	3.91
7	5.14	4.92	4.90
11	7.43	6.72	6.67
16	10.28	8.25	7.86
20	**10.73**	**9.60**	**9.35**
25	10.53	9.58	9.32

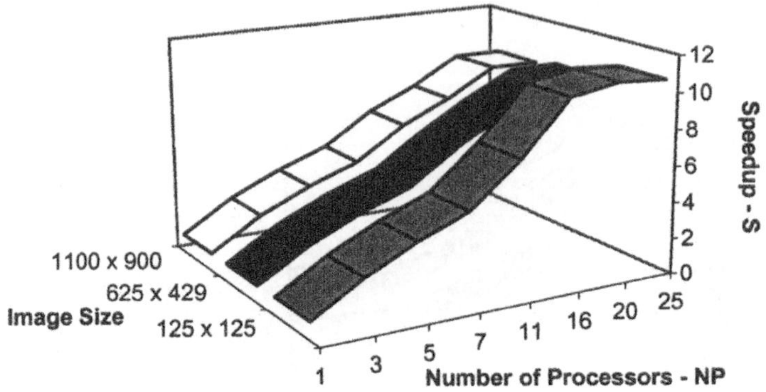

Figure A.2a Speedup graph for Plan P3

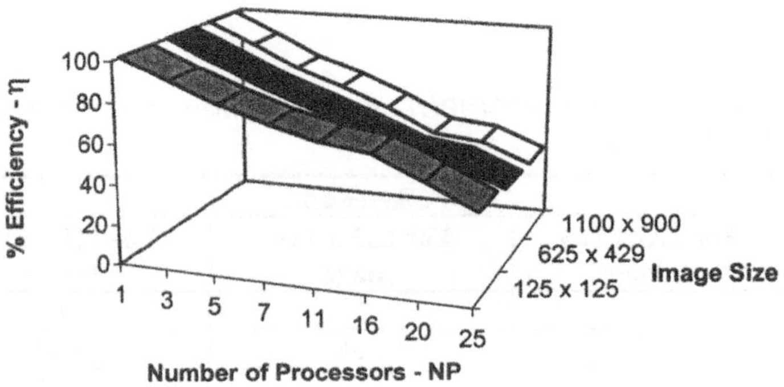

Figure A.2b Efficiency graph for Plan P3

Table A.23Speedup for NIPC Plan P4(NBCT on a Distributed Memory Architecture with Tree Topology)

Number of Processors -NP	Speedup-S		
	For 125 x 125 image	For 625 x 429 image	For 1100 x 900 image
1	1	1	1
3	2.71	2.60	2.60
5	3.42	3.54	3.53
9	4.50	4.56	4.55
15	5.98	5.92	5.92
21	6.75	6.66	6.62
27	**7.25**	**6.99**	**6.97**
33	6.98	6.70	6.89

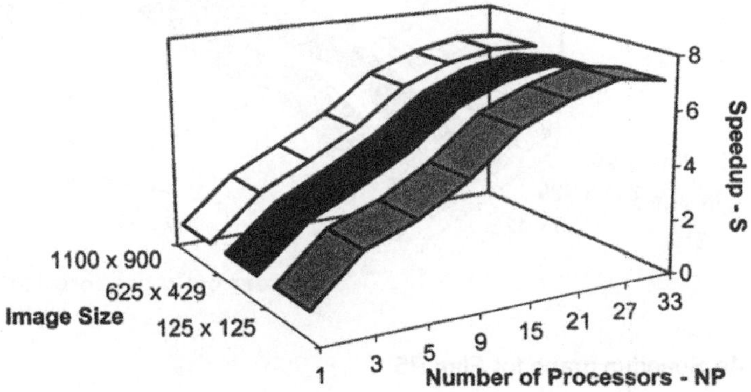

Figure A.3a Speedup graph for Plan P4

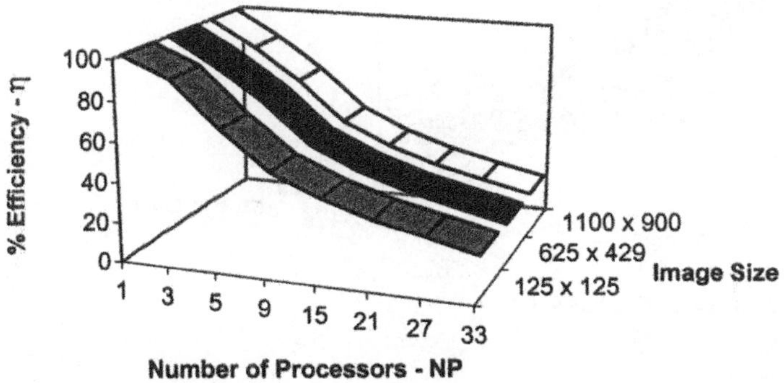

Figure A.3b Efficiency graph for Plan P4

Table A.24Speedup for NIPC Plan P5(NBCT on a Distributed Memory Architecture with Torus Topology)

Number of Processors -NP	Speedup-S		
	For 125 x 125 image	For 625 x 429 image	For 1100 x 900 image
1	1	1	1
3	2.71	2.60	2.60
5	3.42	3.54	3.53
13	5.73	5.79	5.80
17	6.95	6.89	6.87
21	**9.61**	**7.13**	**7.13**
26	6.98	7.03	7.01

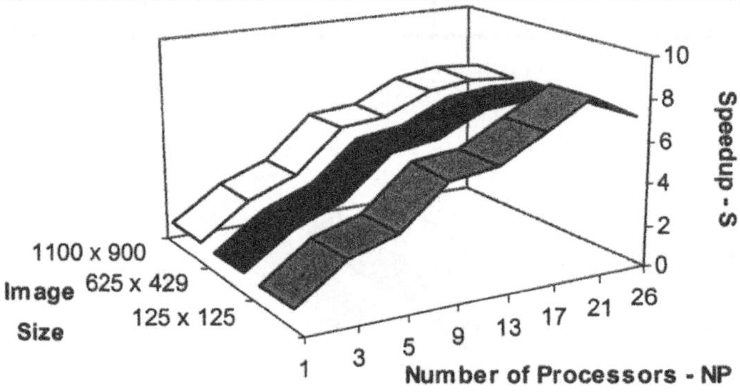

Figure A.4a Speedup graph for Plan P5

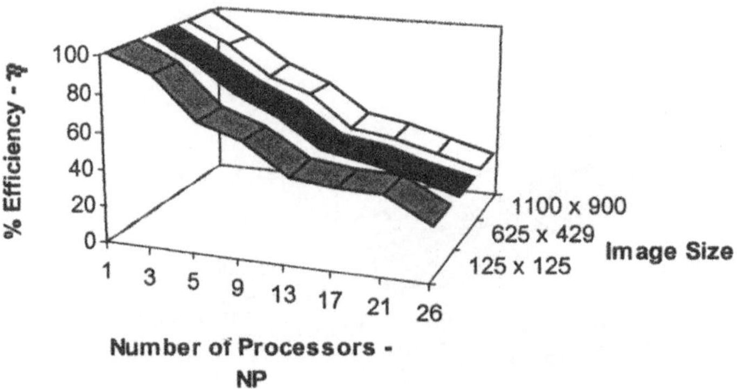

Figure A.4b Efficiency graph for Plan P5

Table A.25Speedup for NIPC Plan P6(NBCT on a Distributed Memory Architecture with Pyramid Topology)

Number of Processors -NP	Speedup-S		
	For 125 x 125 image	For 625 x 429 image	For 1100 x 900 image
1	1	1	1
5	4.81	4.85	4.84
9	5.14	5.20	5.19
21	6.19	6.06	6.07
37	**8.36**	**8.34**	**8.43**
53	7.69	7.80	7.81

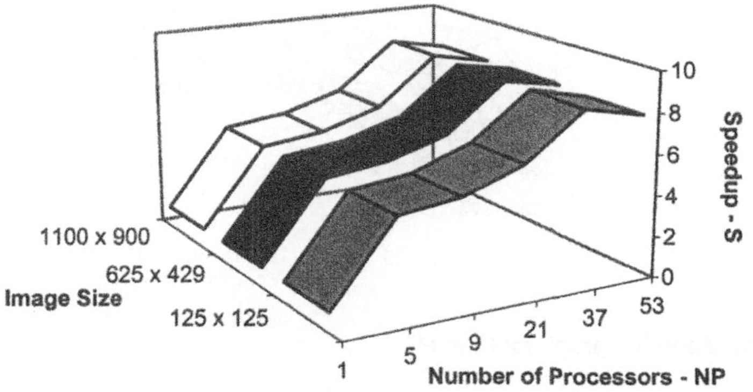

Figure A.5a Speedup Graph for Plan P6

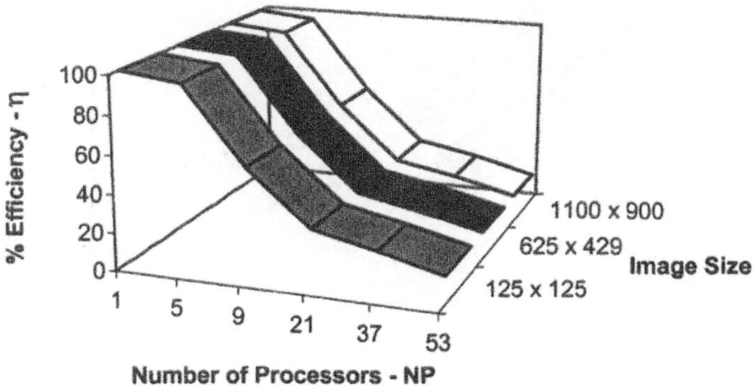

Figure A.5b Efficiency Graph for Plan P6

Table A.26Speedup for NIPC Plan P7(NBCT on a Distributed Memory Architecture with Cube Topology)

Number of Processors -NP	Speedup-S		
	For 125 x 125 image	For 625 x 429 image	For 1100 x 900 image
1	1	1	1
5	4.81	4.85	4.84
9	5.14	5.20	5.19
28	**6.89**	**6.75**	**7.14**
49	3.82	4.83	4.89

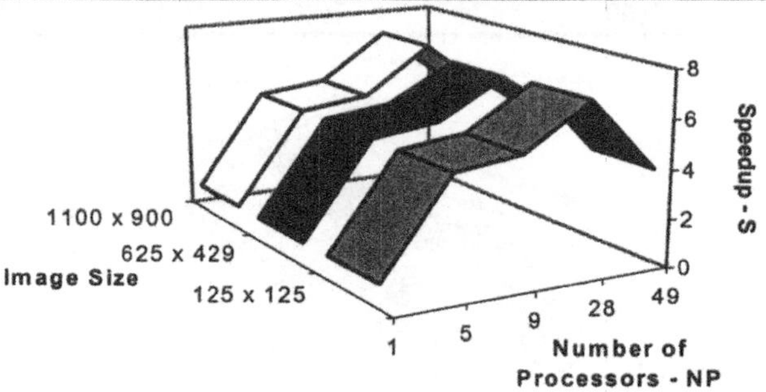

Figure A.6a Speedup graph for Plan P7

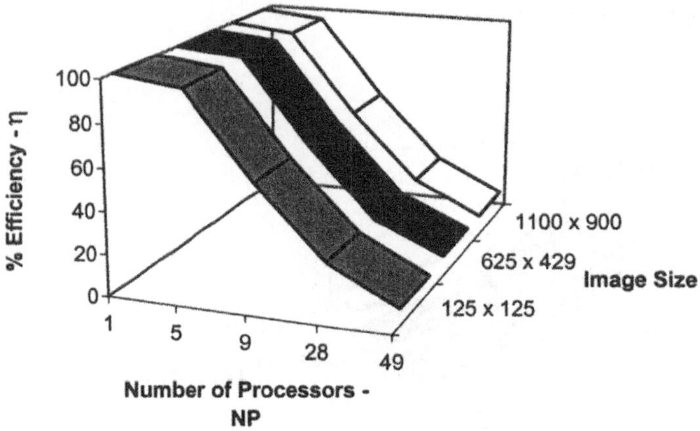

Figure A.6b Efficiency graph for Plan P7

Table A.27Speedup for NIPC Plan P8 (NSB = 10%) (BCT on a Shared Memory Architecture with Global Memory)

Number of Processors -NP	Speedup-S		
	For 125 x 125 image	For 625 x 429 image	For 1100 x 900 image
1	1	1	1
3	2.20	2.43	2.43
5	3.49	3.49	3.49
11	5.52	5.41	5.41
16	**6.30**	**6.33**	**6.35**
20	5.74	5.75	5.75

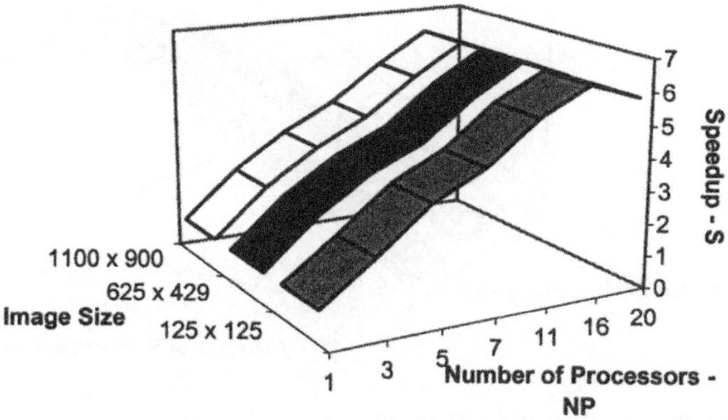

Figure A.7a Speedup graph for Plan P8

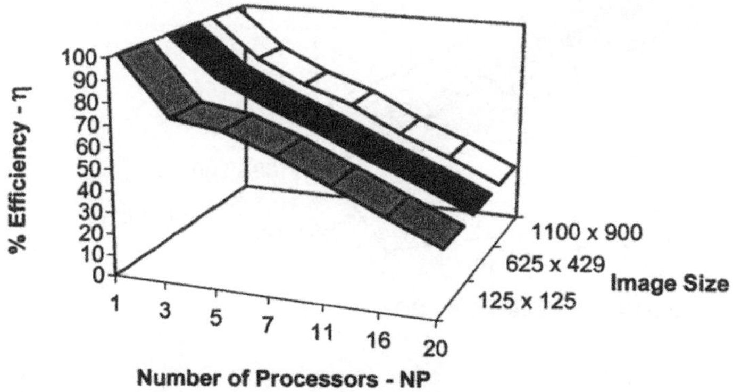

Figure A.7b Efficiency graph for Plan P8

Table A.28Speedup for NIPC Plan P9 (NSB = 10%)(BCT on a Shared Memory Architecture with Local-plus-Global Memory)

Number of Processors-NP	Speedup-S		
	For 125 x 125 image	For 625 x 429 image	For 1100 x 900 image
1	1	1	1
3	2.66	2.65	2.64
5	3.96	3.96	3.91
11	6.85	6.69	6.69
16	8.20	7.62	7.97
25	9.73	8.01	7.80

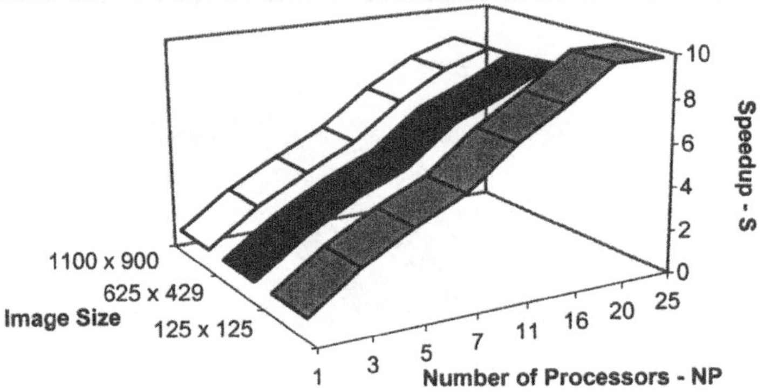

Figure A.8a Speedup graph for Plan P9

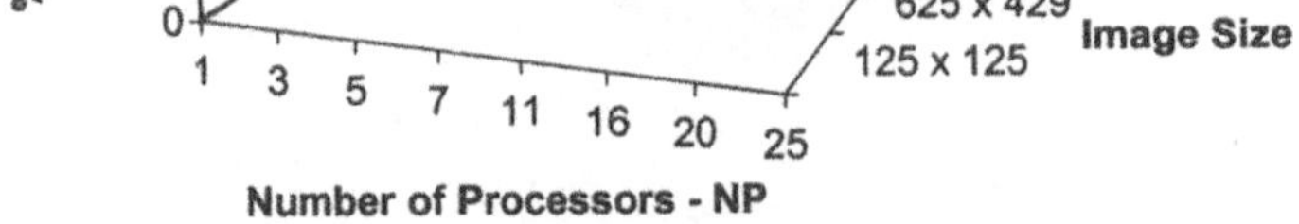

Figure A.8b Efficiency graph for Plan P9

Table A.29 Speedup for NIPC Plan P1 (NSB = 10%)(BCT on a Distributed Memory Architecture with Tree Topology)

Number of Processors-NP	Speedup-S		
	For 125 x 125 image	For 625 x 429 image	For 1100 x 900 image
1	1	1	1
3	2.55	2.53	2.53
5	4.12	3.89	3.88
9	5.25	5.30	4.87
15	**6.56**	**6.72**	**6.19**
21	6.21	6.21	5.75

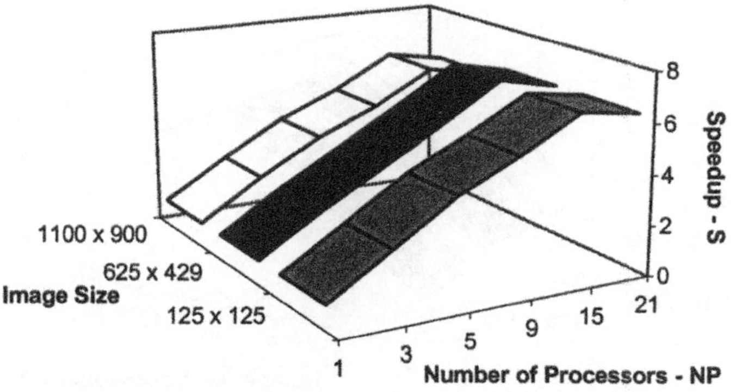

Figure A.9a Speedup graph for Plan P1

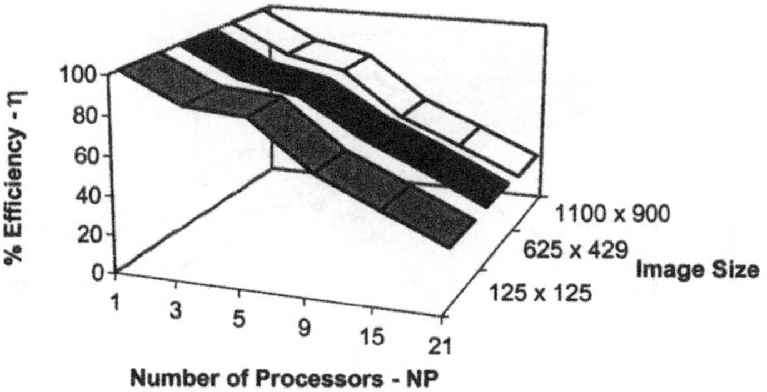

Figure A.9b Efficiency graph for Plan P1

Table A.30Speedup for NIPC Plan P10 (NSB = 10%) (BCT on a Distributed Memory Architecture with Torus Topology)

Number of Processors-NP	Speedup-S		
	For 125 x 125 image	For 625 x 429 image	For 1100 x 900 image
1	1	1	1
3	2.55	2.53	2.53
5	4.12	3.89	3.88
9	**4.91**	**5.01**	4.65
13	4.40	4.36	**5.44**
17			4.81

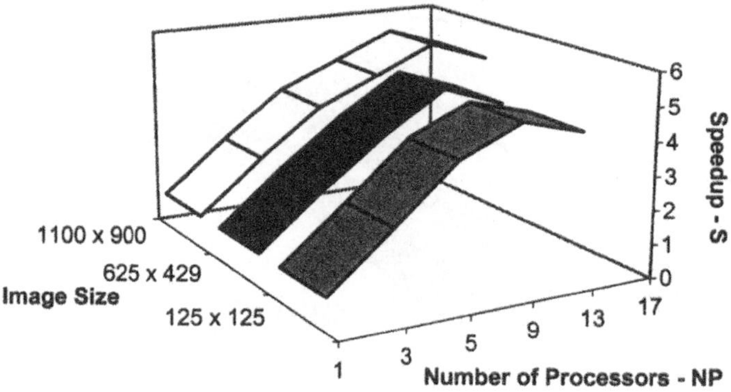

Figure A.10a Speedup graph for Plan P10

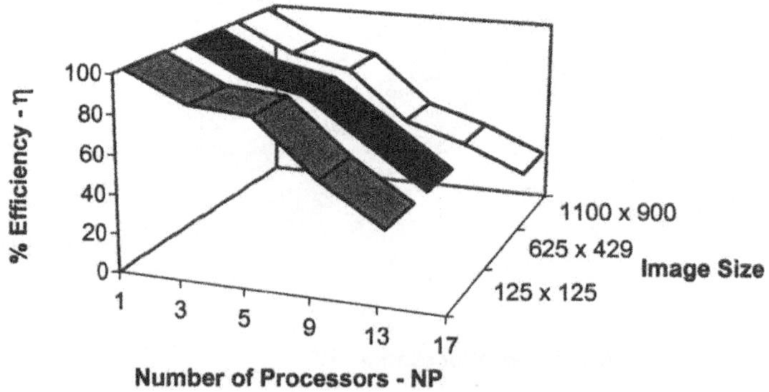

Figure A.10b Efficiency graph for Plan P10

Table A.31Speedup for NIPC Plan P11 (NSB = 10%)(BCT on a Distributed Memory Architecture with Pyramid Topology)

Number of Processors-NP	Speedup-S		
	For 125 x 125 image	For 625 x 429 image	For 1100 x 900 image
1	1	1	1
5	4.70	4.68	4.67
9	5.43	5.48	5.48
21	**7.70**	**7.71**	**7.66**
37	7.50	7.29	7.28

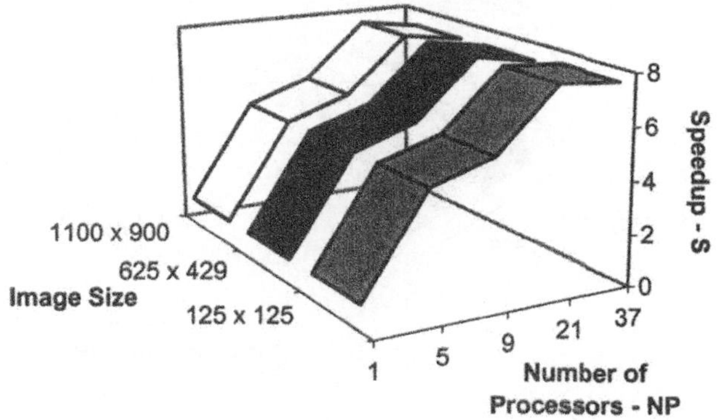

Figure A.11a Speedup graph for Plan P11

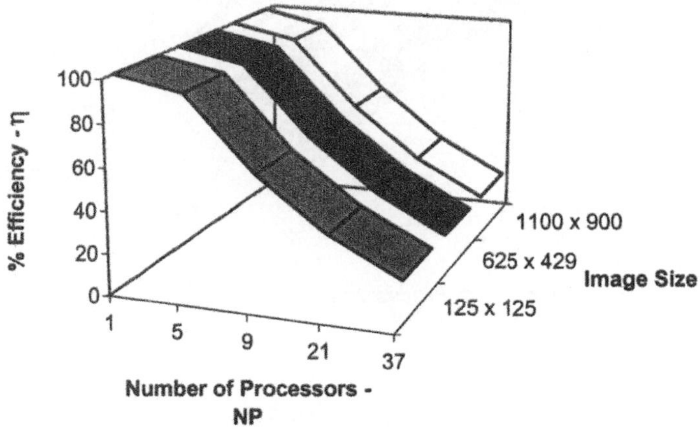

Figure A.11b Efficiency graph for Plan P11

Table A.32Speedup for NIPC Plan P12 (NSB = 10%)(BCT on a Distributed Memory Architecture with Cube Topology)

Number of Processors -NP	Speedup-S		
	For 125 x 125 image	For 625 x 429 image	For 1100 x 900 image
1	1	1	1
5	4.70	4.68	4.67
9	**5.43**	**5.48**	**5.48**
28	5.25	5.21	5.38

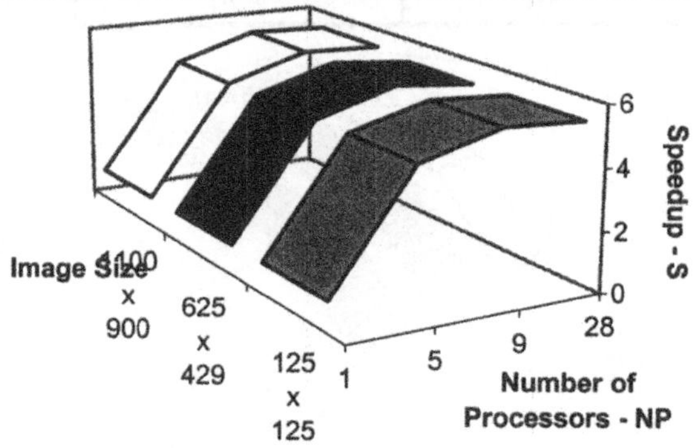

Figure A.12a Speedup graph for Plan P12

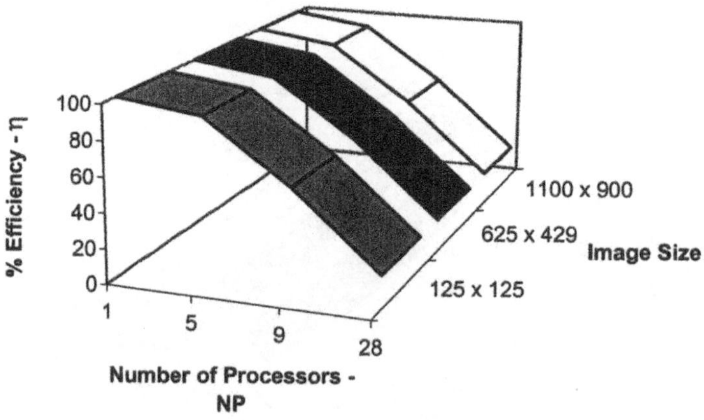

Figure A.12b Efficiency graph for Plan P12

Table A.33Speedup for IPC Plan P13 (NSB = 10%) (BCT on a Shared Memory Architecture with Global Memory)

Number of Processors-NP	Speedup-S		
	For 125 x 125 image	For 625 x 429 image	For 1100 x 900 image
1	1	1	1
5	2.46	3.88	3.86
7	4.03	4.80	4.93
9	4.56	5.30	5.48
13	**5.21**	**5.83**	**5.84**
17	4.84	5.63	5.67

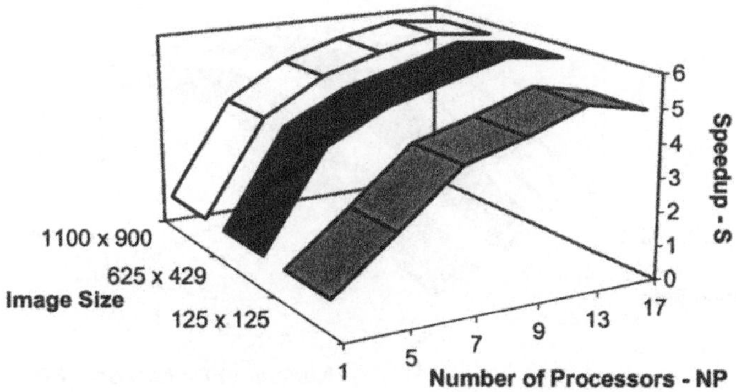

Figure A.13a Speedup graph for Plan P13

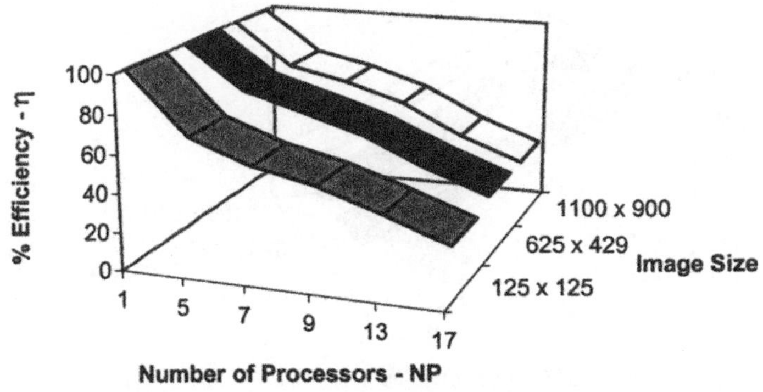

Figure A.13b Efficiency graph for Plan P13

Table A.34Speedup for IPC Plan P14 (NSB = 10%) (BCT on a Distributed Memory Architecture with Torus Topology)

Number of Processors-NP	Speedup-S		
	For 125 x 125 image	For 625 x 429 image	For 1100 x 900 image
1	1	1	1
3	2.47	2.51	2.51
5	3.45	3.97	3.92
9	5.03	5.32	5.58
13	**5.57**	**5.91**	**5.97**
17	5.37	5.65	5.68

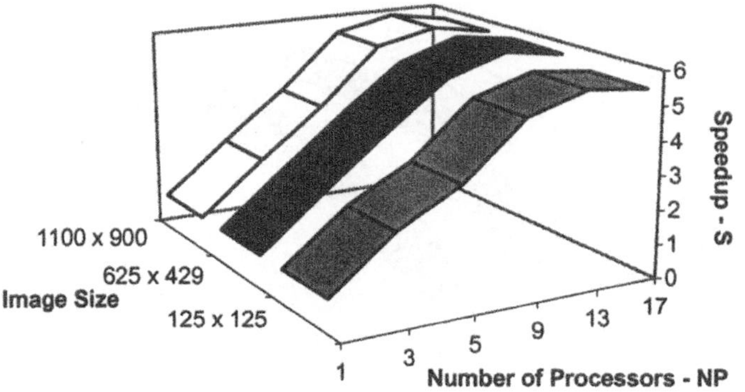

Figure A.14a Speedup graph for Plan P14

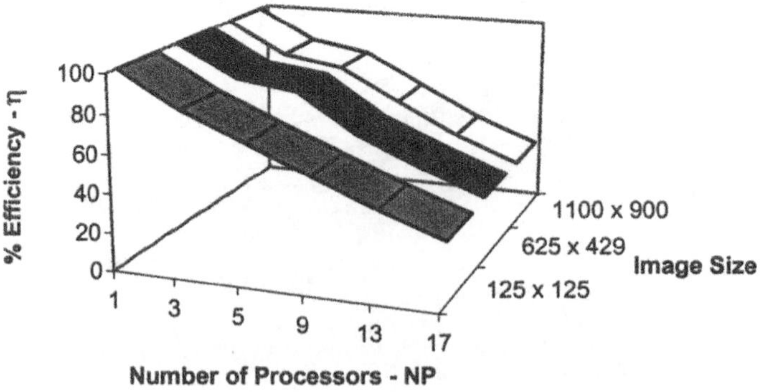

Figure A.14b Efficiency graph for Plan P14

Table A.35Speedup for IPC Plan P15 (NSB = 10%) (BCT on Distributed Memory Architecture with Pyramid Topology)

Number of Processors -NP	Speedup-S		
	For 125 x 125 image	For 625 x 429 image	For 1100 x 900 image
1	1	1	1
5	3.69	3.85	3.87
9	5.19	5.38	6.01
21	**8.70**	**7.93**	**7.69**
37	8.39	7.71	7.38

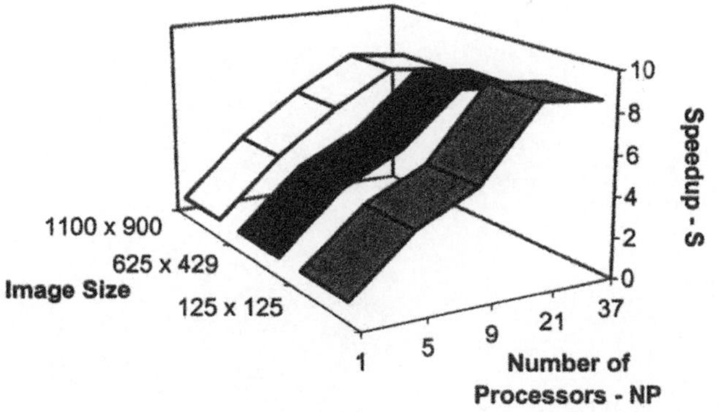

Figure A.15a Speedup graph for Plan P15

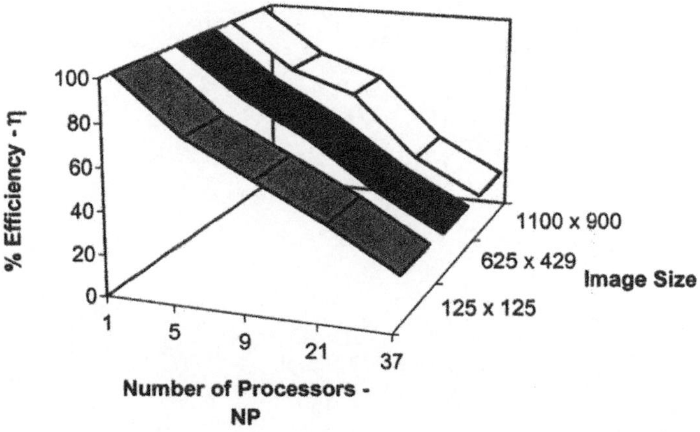

Figure A.15b Efficiency graph for Plan P15

INDEX

M

N

O

P

Q

R

S

T

U

V

W

GPSR Compliance

The European Union's (EU) General Product Safety Regulation (GPSR) is a set of rules that requires consumer products to be safe and our obligations to ensure this.

If you have any concerns about our products, you can contact us on ProductSafety@springernature.com

In case Publisher is established outside the EU, the EU authorized representative is:

Springer Nature Customer Service Center GmbH
Europaplatz 3
69115 Heidelberg, Germany

Batch number: 10755292

Printed by Printforce, the Netherlands